·新·锐·编·程·语·言·集·萃·

Programming in CoffeeScript

CoffeeScript 程序设计

【美】Mark Bates 著

Goddy Zhao 译

人民邮电出版社

北京

图书在版编目（C I P）数据

CoffeeScript程序设计 / (美) 贝茨 (Mark Bates) 著 ; Goddy Zhao译. -- 北京 : 人民邮电出版社, 2013.1
(新锐编程语言集萃)
书名原文: Programming in CoffeeScript
ISBN 978-7-115-30193-2

Ⅰ. ①C… Ⅱ. ①贝… ②G… Ⅲ. ①JAVA语言－程序设计 Ⅳ. ①TP312

中国版本图书馆CIP数据核字(2012)第284974号

内 容 提 要

CoffeeScript 是一门新的编程语言，一门会被编译为 JavaScript 的语言。本书从运行和编译 CoffeeScript 的基础知识开始，逐步介绍其语法、控件结构、函数、集合和类等内容。本书的特色是，通过对相同页面的 CoffeeScript 代码和 JavaScript 代码的直接比较，让读者能够直观地了解 CoffeeScript 是如何改善了 JavaScript 的，进而能够用它构建强大、灵活、可维护、简洁、可靠以及安全的应用程序。此外，作者还在书中给出一些非常宝贵的提示，提醒读者如何才能更好地开发应用程序。

本书是一本理论和实践相结合的 CoffeeScript 入门教材，更是一本能够带领初学者充分理解并快速掌握 CoffeeScript 的好书，非常适合中高级 Web 开发者阅读。

新锐编程语言集萃

CoffeeScript 程序设计

◆ 著　[美] Mark Bates
译　Goddy Zhao
责任编辑　杨海玲

◆ 人民邮电出版社出版发行　北京市崇文区夕照寺街 14 号
邮编　100061　电子邮件　315@ptpress.com.cn
网址　http://www.ptpress.com.cn
三河市海波印务有限公司印刷

◆ 开本：800×1000　1/16
印张：16
字数：359 千字　2013 年 1 月第 1 版
印数：1－3 000 册　2013 年 1 月河北第 1 次印刷

著作权合同登记号　图字：01-2012-6498 号

ISBN 978-7-115-30193-2

定价：45.00 元

读者服务热线：(010)67132692　印装质量热线：(010)67129223
反盗版热线：(010)67171154

版权声明

译者序

随着 Web 2.0 的流行，尤其是如今 Node.js 的诞生，让服务器端 JavaScript 成为了可能，JavaScript 迎来了它生命的第二春。截止至现在，GitHub 中 21%[1]的开源项目都是用 JavaScript 编写的，占据榜首。RedMonk[2]在 2012 年 9 月 12 日发表的一篇关于统计最流行程序设计语言排行的分析报告中显示[3]，JavaScript 是目前最为流行的编程语言。微软公司的 Windows 8 操作系统中，加入了支持直接用 JavaScript 编写本地应用的特性。种种数据和现象都说明，JavaScript 又一次受到了全世界的关注。

然而，没有任何一门编程语言是完美的，JavaScript 也是如此，JavaScript 一直以来都因为其语言本身饱受争议：容易造成污染的全局变量、没有原生的命名空间、一大堆的假值、万恶的 `with` 和 `evil`，等等。为此，Douglas Crockford[4]专门写了一本名为 *JavaScript: The Good Parts*（《JavaScript 语言精粹》）的书，详细介绍了 JavaScript 中的“糟粕”与“精华”。如今，随着 JavaScript 的大规模使用，人们从心底里期待一个更好的 JavaScript。为此，官方的 ECMAScript[5]工作小组也制定了第 5 版的标准，Chrome、Firefox 等浏览器也开始陆续实现并支持第 5 版的新特性。除此之外，民间的声音也愈渐响亮，其中最有代表性的就当属 CoffeeScript 了。

和 Ruby 的到来让人眼前一亮一样，CoffeeScript 也给编写 JavaScript 带来了快感。简洁的语法让人把更多的精力放在逻辑本身，默认生成的匿名包装器函数让人无须再为全局变量的污染提心吊胆，操作符与别名让人无须再为一大堆的假值烦恼，内置的基于类的继承机制让人无须再为了搞清楚原型那些东西而绞尽脑汁。和 Ruby 一样，CoffeeScript 让人们更加快乐地编程。

对于 JavaScript 程序员来说，CoffeeScript 的学习成本非常低，因为它不像 Ruby 那样是一门全新的语言，它更多的是对 JavaScript 的扩展和增强；对于非 JavaScript 程序员而言，CoffeeScript 也很容易掌握，因为它的语法非常简洁，CoffeeScript 官方网站[6]用短短一页篇幅就展示了所有的语法。更何况，还有本书这样一本理论与实践并存、宽度与广度并具的 CoffeeScript 教材，可以帮助程序员更好地学习 CoffeeScript。

我始终坚信学习技术最好的方式就是实践，在实践的过程中遇到问题，再解决问题；在解决问题的过程中再去深入探寻其背后的理论依据，从而体系化地掌握学习的内容。本书就是这样一本理论和实践相结合的 CoffeeScript 入门教材。最为重要的是，它更加侧重于实践，提供

---

① https://github.com/languages
② http://redmonk.com/
③ http://redmonk.com/sogrady/2012/09/12/language-rankings-9-12/
④ http://www.crockford.com/
⑤ http://ecmascript.org
⑥ http://coffeescript.org/

了一个非常完整的包括前后端实现的待办事宜列表应用，带领读者学习和掌握 CoffeeScript。这也是我很喜欢本书的一个重要原因。除此之外，本书也不缺乏详细的理论，花了 4 章左右的篇幅详细介绍了 CoffeeScript 所有的语法和精髓。总的来说，这是一本能够带领初学者充分理解并快速掌握 CoffeeScript 的好书。

另外，我有责任提醒各位读者：CoffeeScript 也有它的不足，其中最为致命的就是没有很好的调试方式。原因就在于，编写用的是 CoffeeScript，而真正运行的是其编译后的 JavaScript 代码，因此，一旦出了错误，很难通过 JavaScript 错误来定位到底 CoffeeScript 源代码中哪里出错了。不过，好在 Chrome 的开发者工具以及 Firefox 的 Firebug 都在尝试添加 CoffeeScript 的调试功能，相信不久的将来，这个问题会得到解决。

最后，祝愿所有的读者都能够通过本书学习和掌握 CoffeeScript，与此同时，也能体会到编程的乐趣。

前　言

1999 年，我开始了专业的开发生涯，那年我第一次以开发者的身份领取薪水（那年是我开始对 Web 开发产生浓厚兴趣的时候，我并未将此前的几年开发时间计算在内。）。1999 年，Web 还是个“险恶之地”，HTML 中最常见到的还是 `font` 和 `table` 标签，CSS 初出茅庐，JavaScript [1]也才刚刚崭露头角，并且主流浏览器之间还弥漫着对 JavaScript 各自不同实现的硝烟。这就意味着，编写的某段 JavaScript 代码可以在某个浏览器上正常工作，但却不一定能在其他浏览器上正确运行。正因如此，JavaScript 在 21 世纪初名声并不好。

在 21 世纪中期发生了两件重要的事情，改变了开发者对 JavaScript 的看法。第一件就是 AJAX[2]。它能让开发人员制作出交互性更好、速度更快的 Web 页面，使得用户无须手动刷新浏览器就能在后台对服务器进行远程调用。

第二件重要的事情就是 JavaScript 库的流行，如 Prototype[3]，它可以简化跨浏览器的 JavaScript 开发。AJAX 提高了 Web 应用的用户体验，像 Prototype 这样的 JavaScript 库则保证了 Web 应用可以跨主流浏览器工作。

到了 2010 年，当然也包括 2011 年，Web 应用逐步演变为“单页面”应用（single page application）。这类应用使用像 Backbone.js[4]这样的 JavaScript 框架。这类框架用 JavaScript 实现了前端的 MVC[5]设计模式供开发者使用。整个应用使用 JavaScript 进行构建，在终端用户的浏览器端进行载入和运行。这些都服务于构建高响应、富客户端应用。

然而，对开发者而言，这并非就尽善尽美。尽管框架和工具简化了这类应用的开发，但是众所周知，JavaScript 语言本身有很多诟病，它是一门让人喜忧参半的语言。喜的是它非常强大，忧的是它充满了矛盾和设计陷阱，很快会让你的代码变得难以管理，而且会有隐蔽的潜在 bug。

开发者该如何是好？他们想要开发此类新应用，但是唯一被接受的浏览器语言就是 JavaScript。当然了，他们可以用 Flash[6]来写这类应用，以避免使用 JavaScript，但是这样一来就需要插件支持，并且在像 iOS[7]这样的不支持 Flash 的平台下就无法工作。

2010 年 10 月，我“邂逅”了 CoffeeScript[8]，当时它扬言要“驯服”JavaScript，将 JavaScript 这门怪异的语言最优秀的部分呈现出来。它拥有更为清晰的语法，比方说：CoffeeScript 摒弃了

---

① http://en.wikipedia.org/wiki/JavaScript
② http://en.wikipedia.org/wiki/Ajax_(programming)
③ http://www.prototypejs.org/
④ http://documentcloud.github.com/backbone/
⑤ http://en.wikipedia.org/wiki/Model–view–controller
⑥ http://www.adobe.com/
⑦ http://www.apple.com/ios/
⑧ http://www.coffeescript.org

大部分的标点符号[①]，采用了有意义的空格，从而保护 JavaScript 开发者免遭 JavaScript 语言设计缺陷产生的陷阱，例如糟糕的作用域问题以及比较操作符的误用。其中最值得称赞的是，以上这些工作，CoffeeScript 都会在将源代码编译为标准的 JavaScript 的过程中完成，编译出来的 JavaScript 可以在任意浏览器以及其他的 JavaScript 运行时环境中执行。

在我刚使用 CoffeeScript 时，尽管当时版本号已经是 0.9.4 了，但这门语言本身仍然"很粗糙"，还有不少问题。当时，我将其用在一个项目的客户端部分，主要是想看看它是否真的有我听说的那么好。但遗憾的是，那个时候有两个原因促使我将其搁置。第一个原因是，它还不够"成熟"，还有太多的 bug 并且功能尚未完善。

第二个原因是，我尝试使用 CoffeeScript 的应用并非是一个 JavaScript 很重的应用[②]。此应用中只用到了少量的校验和 AJAX，大部分的工作都让 Ruby on Rails[③]帮我处理掉了。

那么又是什么促使我"重拾"CoffeeScript 的呢？在我首次尝试 CoffeeScript 之后——6 个月之后的某一天，官方宣布[④]Rails 3.1 将以 CoffeeScript 作为其默认的 JavaScript 引擎。和绝大多数开发者一样，我被这个消息吸引了。要知道我之前就用过 CoffeeScript 并且觉得它不怎么好。他们是怎么想的呢？

和其他开发同事不同的是，我又花时间看了一下 CoffeeScript。6 个月对于任何一个项目来说都是很长的一段开发周期，CoffeeScript 也走过了相当长的一段路，因此我决定再次尝试 CoffeeScript，这次将它运用在 JavaScript 代码占相当大比重的应用中。经过几天的使用，我不仅改变了对 CoffeeScript 的看法，甚至成为了它的拥护者。

我不打算告诉你究竟是什么改变了我，也不想告诉你我为什么喜欢它。我想让你做出自己的选择。我希望通过阅读本书不仅能够改变你，而且也能让你成为这门漂亮、小巧的语言的拥护者，而这一切，都是出于你本人的意愿。不过，我会带你初窥一下后面的内容。下面是一段摘自真实应用的 CoffeeScript 脚本，其后是等效的 JavaScript 脚本。尽情享受吧！

例（源代码：sneak_peak.coffee）

```
@updateAvatars = ->
  names = $('.avatar[data-name]').map -> $(this).data('name')
  Utils.findAvatar(name) for name in $.unique(names)
```

例（源代码：sneak_peak.js）

```
(function() {

  this.updateAvatars = function() {
  var name, names, _i, _len, _ref, _results;
  names = $('.avatar[data-name]').map(function() {
    return $(this).data('name');
```

① 比方说，最常见的分号、大括号等。——译者注
② 即富客户端应用。——译者注
③ http://www.rubyonrails.org
④ http://www.rubyinside.com/rails-3-1-adopts-coffeescript-jquery-sass-and- controversy-4669.html

```
    });
    _ref = $.unique(names);
    _results = [];
    for (_i = 0, _len = _ref.length; _i < _len; _i++) {
      name = _ref[_i];
      _results.push(Utils.findAvatar(name));
    }
    return _results;
  };

}).call(this);
```

什么是 CoffeeScript

CoffeeScript 是一门会被编译为 JavaScript 的语言。我知道这个解释的内容并不丰富，但是事实确实如此。CoffeeScript 和 Ruby①以及 Python②很像，它设计的初衷就是帮助开发者更加高效地编写 JavaScript 代码。通过移除没有必要的诸如括号、分号这样的标点符号，用有意义的空格取而代之，可以让开发者更多地将注意力集中在代码本身，而不必去关心大括号是否关闭这样的问题。

比方说，你可能会写下面这样的 JavaScript 代码。

例（源代码：punctuation.js）

```
(function() {

  if (something === something_else) {
    console.log('do something');
  } else {
    console.log('do something else');
  }

}).call(this);
```

那么为什么不试着把它写成下面的形式。

例（源代码：punctuation.coffee）

```
if something is something_else
  console.log 'do something'
else
  console.log 'do something else'
```

CoffeeScript 还提供了几种快捷方式来简化复杂代码块的编写。例如，下面这段代码就可以实现数组值循环，不需要关心其下标。

① http://en.wikipedia.org/wiki/Ruby_(programming_language)
② http://en.wikipedia.org/wiki/Python_(programming_language)

例 （源代码: array.coffee）

```
for name in array
  console.log name
```

例 （源代码: array.js）

```
(function() {
  var name, _i, _len;

  for (_i = 0, _len = array.length; _i < _len; _i++) {
    name = array[_i];
    console.log(name);
  }

}).call(this);
```

除了这些改进的语法糖之外，CoffeeScript 还可以帮助写出更好的 JavaScript 代码。例如，帮助准确地确定变量和类的作用域，帮助确保正确使用了比较操作符，等等。这些在阅读本书时都能看到。

CoffeeScript、Ruby 以及 Python 经常会因它们相似的优点共同受到关注。CoffeeScript 直接借鉴了 Ruby、Python 这样的语言提供的简洁语法。正因如此，CoffeeScript 比 JavaScript 这门更像 Java[①]或者 C++[②]的语言更具"现代感"。与 JavaScript 相同的是，CoffeeScript 可以在任何编程环境中使用。不论是用 Ruby、Python、PHP[③]、Java 还是用.NET[④]开发的应用，都没有关系。编译后的 JavaScript 代码都可以很好地和它们一起工作。

因为 CoffeeScript 会编译为 JavaScript，所以你仍然可以使用现在使用的所有 JavaScript 库，可以用 jQuery[⑤]、Zepto[⑥]、Backbone[⑦]、Jasmine[⑧]等，并且它们都可以很好地工作。而这种情况对于一门新语言来讲并不多见。

听起来很不错吧，不过我仿佛听到了你在问：与 JavaScript 相比，CoffeeScript 的缺点有哪些呢？这是个很好的问题。答案是：缺点肯定有，但还不算多。首先，尽管 CoffeeScript 是一种很好的编写 JavaScript 代码的方式，但是，凡是 JavaScript 无法实现的，使用 CoffeeScript 也依然无能为力。比方说，我不可能用 CoffeeScript 创建一个类似 Ruby 中最著名的 method_missing[⑨]的 JavaScript 版本。另外，CoffeeScript 的最大缺点就是，对于你和你的团队成员，它完全是另外一门要学习的语言。不过好在，它足够简单，你会看到，CoffeeScript 非常容易学。

最后，如果出于某种原因，CoffeeScript 不适合你或者你的项目，你也可以使用生成的 JavaScript。总之，说真的，你没有理由不在下一个项目甚至是当前项目中尝试 CoffeeScript

① http://en.wikipedia.org/wiki/Java_(programming_language)
② http://en.wikipedia.org/wiki/C%2B%2B
③ http://en.wikipedia.org/wiki/Php
④ http://en.wikipedia.org/wiki/.NET_Framework
⑤ http://www.jquery.com
⑥ https://github.com/madrobby/zepto
⑦ http://documentcloud.github.com/backbone
⑧ http://pivotal.github.com/jasmine/
⑨ http://ruby-doc.org/docs/ProgrammingRuby/html/ref_c_object.html#Object.method_ missing

（CoffeeScript 和 JavaScript 互相配合得很好）。

本书适合什么样的读者

本书适合中高级 JavaScript 开发者。之所以我觉得本书不大适合对 JavaScript 不熟悉或者仅对其略知一二的人是有原因的。

首先，本书不会教 JavaScript，这是一本介绍 CoffeeScript 的书。虽然在读的过程中你肯定也能学到一些零碎的 JavaScript 知识（CoffeeScript 会让你学到更多关于 JavaScript 的知识），但是本书不会从头开始教你 JavaScript。

例　这段代码做了什么？（源代码：example.js）

```
(function() {
  var array, index, _i, _len;

  array = [1, 2, 3, 4, 5, 6, 7, 8, 9, 10];

  for (_i = 0, _len = array.length; _i < _len; _i++) {
    index = array[_i];
    console.log(index);
  }

}).call(this);
```

如果你不知道上面这段示例代码是什么意思，我建议你就此打住。不用担心，我是真心想要你再回来继续阅读下去。我只是觉得，如果你能对 JavaScript 有深入的理解，就能最大限度地利用这本书。在本书的讲述过程中，我通常会介绍一些 JavaScript 的基础知识，帮助你更好地理解 CoffeeScript。不过，尽管如此，在继续阅读之前，对 JavaScript 有一定的基础还是非常重要的。因此，请去找本介绍 JavaScript 的优秀书籍（有很多这样的书）来读，然后再和我一起踏上成为 CoffeeScript 高手的旅途。

对于已经是 JavaScript 高手的你，让我们一起踏上旅途吧！本书将教你如何使用 CoffeeScript 编写更简洁、更好的 JavaScript 代码。

如何阅读本书

这里我提供一些阅读本书的方法，希望能帮助你循序渐进地学习 CoffeeScript。第一部分中的各章应该按顺序阅读，因为每一章都是建立在前一章的基础之上的，所以请不要跳着阅读。

在阅读每一章的过程中，你会注意到以下一些事情。

首先，每当我介绍一些外部库、想法和概念的时候，都会加上脚注，写上相应的网站地址，通过它你可以学到更多相关的内容。尽管我很想和你喋喋不休地讲些其他东西，像 Ruby，但是本书篇幅有限。因此，如果对我提到的某些东西感兴趣，想要在继续阅读之前了解更多相关的信息，那就请你访问对应的网站，满足你对知识的渴望之后，再回来继续阅读本书。

其次，在每一章中，我有时会先介绍问题错误的解决方案。在看明白了错误的方式后，我们才能审视它、理解它，进而想出正确的解决方案。运用这种方式的一个典型例子在第 1 章中，我们讨论了从 CoffeeScript 编译为 JavaScript 的多种不同方式。

本书中还会出现如下内容：

> **提示**：这里是一些有用的提示。这些是我认为或许会对你有帮助的一些提示。

最后，本书中通常都会一次出现两三段代码块。首先是 CoffeeScript 脚本，然后是对应的编译后的 JavaScript 脚本，如果示例有输出的话，最后，可能还会有示例的输出结果（如果我认为有必要展示的话），如下所示。

例 （源代码：example.coffee）

```
array = [1..10]

for index in array
  console.log index
```

例 （源代码：example.js）

```
(function() {
  var array, index, _i, _len;

  array = [1, 2, 3, 4, 5, 6, 7, 8, 9, 10];

  for (_i = 0, _len = array.length; _i < _len; _i++) {
    index = array[_i];
    console.log(index);
  }

}).call(this);
```

输出 （源代码：example.coffee）

```
1
2
3
4
5
6
7
8
9
10
```

有时，还会特意展示一些错误，如下所示。

例 （源代码：oops.coffee）

```
array = [1..10]

oops! index in array
  console.log index
```

输出 （源代码：oops.coffee）

```
Error: In content/preface/oops.coffee, Parse error on line 3: Unexpected 'UNARY'
    at Object.parseError (/usr/local/lib/node_modules/coffee-script/lib/coffee-script/
➥parser.js:470:11)
    at Object.parse (/usr/local/lib/node_modules/coffee-script/lib/coffee-script/
➥parser.js:546:22)
    at /usr/local/lib/node_modules/coffee-script/lib/coffee-script/coffee-script.
➥js:40:22
    at Object.run (/usr/local/lib/node_modules/coffee-script/lib/coffee-script/
➥coffee-script.js:68:34)
    at /usr/local/lib/node_modules/coffee-script/lib/coffee-script/command.js:135:29
    at /usr/local/lib/node_modules/coffee-script/lib/coffee-script/command.js:110:18
    at [object Object].<anonymous> (fs.js:114:5)
    at [object Object].emit (events.js:64:17)
    at afterRead (fs.js:1081:12)
    at Object.wrapper [as oncomplete] (fs.js:252:17)
```

本书的组织结构

为了让读者从本书中尽可能地获益，我将本书分为两个部分。

第一部分：核心 CoffeeScript

本书第一部分自上向下地介绍了整个 CoffeeScript 语言。阅读完这部分内容之后，就完全可以应对各种 CoffeeScript 项目了，包括第二部分内容里提到的那些。

第 1 章“从这里开始”，介绍各种编译和运行 CoffeeScript 的方法。除此之外，这一章还会介绍 CoffeeScript 自带的功能强大的命令行工具和 REPL。

第 2 章“基础知识”，介绍 CoffeeScript 与 JavaScript 的不同点。内容涵盖语法、变量、作用域，以及更多后续章节所需的基础知识。

第 3 章“控制结构”，介绍语言中的一个重要部分——控制结构，比如 `if`、`else`。这一章还会介绍 CoffeeScript 中的操作符和 JavaScript 中的操作符的区别。

第 4 章“函数与参数”，详细介绍 CoffeeScript 中的函数。内容涵盖函数定义、函数调用以及一些额外的东西，比如默认参数和 splat 操作符。

第5章“集合与迭代”，从数组到对象，依次介绍在CoffeeScript如何使用、操作，以及迭代集合对象。

第6章“类”，第一部分的最后一章，介绍CoffeeScript中的类。内容涵盖类的定义、类的扩展、函数重载等。

第二部分：CoffeeScript实践

本书第二部分介绍CoffeeScript实践。通过学习CoffeeScript周边的生态系统以及构建完整的应用，让你精通CoffeeScript。

第7章“Cake与Cakefile”，介绍CoffeeScript自带的Cake工具。它用来创建构建脚本、测试脚本等。与之相关的概念都会介绍。

第8章“使用Jasmine测试”，测试是软件开发过程中非常重要的一环，这一章会带你快速一览最流行的CoffeeScript/JavaScript测试库之一——Jasmine。这一章还会通过一个计算器类的测试脚本来带着你体验一把流行的测试驱动开发模式。

第9章“Node.js介绍”，简单介绍事件驱动的服务器端框架——Node.js。这一章会使用CoffeeScript来构建一个简单的HTTP服务器，该服务器可以根据网页浏览器的请求自动将CoffeeScript文件编译为JavaScript文件。

第10章“示例：待办事宜列表第1部分（服务器端）”，将介绍构建一个待办事宜列表应用的服务器端部分。在第9章的基础上，我们用Express.js web框架以及MongoDB的ORM框架Mongoose来构建一个API。

第11章“示例：待办事宜列表第2部分（客户端，使用jQuery）”，介绍使用流行的jQuery库来为第10章中建立的待办事宜列表API构建客户端部分。

第12章“示例：待办事宜列表第3部分（客户端，使用Backbone.js）”，抛弃jQuery，采用客户端框架Backbone.js对待办事宜列表应用进行重构。

安装CoffeeScript

我不是很喜欢在书中介绍安装指南，主要是因为到了书上架开售的时候，这些安装指令也许已经过时了。不过，有的时候出版社编辑会觉得应当在书中介绍安装指南。本书便是如此。

安装CoffeeScript非常简单。最简单的方式就是访问CoffeeScript的官方网站，网址是http://www.coffeescript.org/，然后根据上面的安装指南安装即可。

我相信像CoffeeScript和Node[①]这样的项目的维护者一定会保证官方网站上的安装指南的时效性，所以，直接根据官网的安装指南安装是最好的方式。

① http://nodejs.org

本书撰写时，CoffeeScript 的版本号是 1.2.0。书中所有的例子在该版本的 CoffeeScript 下都可以工作。

如何运行书中示例

在 https://github.com/markbates/Programming-In-CoffeeScript 可以下载本书所有示例的源代码。该站点上的内容一目了然，所有的示例都可以在对应的示例文件中找到。示例文件可以根据出现的章号在对应的章文件夹中找到。

除非有特殊说明，否则，所有的示例代码应该都能在终端以如下形式运行：

```
> coffee example.coffee
```

现在你已经知道如何运行本书示例代码了，再安装好 CoffeeScript，我们就可以马上进入第 1 章的内容了。

致 谢

有句话在我第一本书中已经说过，在这里我还要重复一遍：写书真的是一项艰苦的工作！我相信不会有人对此有不同的感受。如果真的有，要么他在撒谎，要么他就是斯蒂芬·金[①]。幸运的是，我介于两者之间。

写书既是一项独立的工作，也需要团队的努力。我哄孩子睡觉后，就一头扎进了书房，一连打开几瓶吉尼斯黑啤，调大音乐声，开始独自写书，一连好几个小时，一直到凌晨。每当完成一章，就发给我的编辑，他会把我的稿子发给其他一些人从不同方面对其进行改进，这些人我可能都不认识。他们简单的会纠正语法或者拼写错误，复杂的会帮助改善书的结构或者指出书中示例代码不清楚的地方。因此，说真的，写作可能是一个人独自在小黑屋中完成的，但是，最终的产品一定是一群人共同努力的结果。

在这里，我有机会对所有为你现在手中（或者下载的）这本高质量的书做出过努力的人表示感谢。下面，请允许我效仿奥斯卡颁奖礼致感谢词，对在感谢名单上遗漏的人，我深感抱歉。

首先，我最要感谢的是我美丽的太太Rachel。她是我见过的最支持我也最坚强的人之一。每一个夜晚在她身边入睡，每一个早晨从她身旁醒来。凝视着她的双眸，我能看到无私的爱，无比幸福。而且，不论我写书、创业还是做任何我觉得开心的事情，她都在背后支持我。她给了我两个帅气的儿子，反过来，我给她的则是我蹩脚的幽默和我用过的手机。我俩的婚姻中，我无疑是受益者，为此，我会永远心存感激。

接下来，我要感谢我的两个儿子 Dylan 和 Leo。尽管他们没有对本书作出直接贡献，但是他们给了我对生活的动力和激情，这一切除了孩子无人能给。儿子，爸爸非常爱你们。

我还要感谢我的父母（特别是你，母亲！）以及其他的家庭成员，谢谢你们一直以来对我的支持，同时，还时刻让我保持自省。我爱你们。

下面，我还要感谢 Debra Williams Cauley。Debra 是本书的编辑、负责人，同时也是我第一本书——《Ruby 分布式编程》（*Distributed Programming with Ruby*）的“心理医生”。我只希望其他的作者也能同样幸运，能有机会与像 Debra 这样好的编辑合作。她真的非常有耐心。

我希望下次再写书的话还能与 Debra 合作。我无法想象写书要是没有她会怎样。真的谢谢你，Debra。

在写技术书的过程中，有些人至关重要，那就是技术审校者。技术审校者的工作就是阅读每个章节，并从技术角度对内容进行评论，就好像在回答这样的问题：“这里介绍这些合适吗？”

---

① 斯蒂芬·金是一位作品多产，屡获奖项的美国畅销书作家，编写过剧本、专栏评论，曾担任电影导演、制片人以及演员。——译者注

他们扮演书的读者，同时又都懂技术。因此，他们的反馈非常重要。本书有几位审校者，其中我特别想提的两位是 Stuart Garner 和 Dan Pickett。他们对审校工作非常负责，甚至做了许多超越其职责的事情，而且对我做过的蠢事或者说过的蠢话直言不讳。他们总要被我不分昼夜的邮件和电话打扰，但却总能给我很好的反馈。要不是我想“独吞”版税，我真想把版税分给他们。（别担心，他们也会得到自己的报酬。他们每个人都得到了工作时间一小时的休息。）谢谢 Dan、Stuart 以及其他审校者们，谢谢你们的辛苦付出。

我要感谢许多以不同方式对本书做出贡献的朋友。有人为本书设计了封面，有人建了索引，有人实现了程序设计语言（CoffeeScript）以及其他与此相关的数之不尽的工作。下面是这些人（我知道的）的名单，排名不分先后： Jeremey Ashkenas、Trevor Burnham、Dan Fishman、Chris Zahn、Gregg Pollack、Gary Adair、Sandra Schroeder、Obie Fernandez、Kristy Hart、Andy Beaster、Barbara Hacha、Tim Wright、Debbie Williams、Brian France、Vanessa Evans、Dan Scherf、Gary Adair、Nonie Ratcliff 以及 Kim Boedigheimer。

我还要感谢所有在我开始写书时听我唠叨的人。我知道，这对绝大多数人而言有点儿枯燥，但是，我就是喜欢听自己的声音。感谢所有容忍我啰唆的朋友。

最后，我要感谢所有读者。感谢你们购买本书，并支持像我这样的人，纯粹只想将自己的知识分享给别人来帮助开发者们。为了你们，我投入了大量的工作时间以及精力来书写本书。我希望在你们阅读完本书时，能够对 CoffeeScript 有更深的理解，并希望它能够对你们的开发工作产生影响。祝你好运！

目　录

第一部分　核心 CoffeeScript

第二部分 CoffeeScript 实践

第一部分　核心 CoffeeScript

本书第一部分涵盖所有你想知道的以及需要知道的 CoffeeScript 知识。读完第一部分就意味着你已经做好进入 CoffeeScript 实战的准备，同时也熟悉了它提供的工具，并且对 CoffeeScript 语言本身有了充分的理解。

我们将先从最基础的内容开始，如学习如何编译和执行 CoffeeScript 文件；然后逐步开始学习 CoffeeScript 的语法。在习惯了它的语法之后，紧接着我会介绍控制结构、函数、集合，最后介绍类。

每章都是建立在上一章的基础上的。“一路上”，你会学到所有 CoffeeScript 拥有的优秀的技巧，来帮助你书写出出色的基于 JavaScript 的应用。还等什么，内容丰富 —— 让我们开始吧！

第 1 章

从这里开始

至此，你已阅读过前言部分，也安装了 CoffeeScript，那我们该如何使用 CoffeeScript 呢？本章开始介绍几种编译和执行 CoffeeScript 的方式。

我打算介绍几种好的编译和执行 CoffeeScript 的方式，与此同时，也会介绍一些不好的方式。尽管本章不会介绍 CoffeeScript 的内部工作机制，但对于刚刚开始使用 CoffeeScript 的人来说，本章绝对是不容错过的。熟练掌握 CoffeeScript 自带的命令行工具，不仅有助于阅读此书，更有助于开发第一个 CoffeeScript 应用。

哪怕你已经玩转了 CoffeeScript 的命令行工具，本章还是可能有你不了解的内容，因此，别急着跳去阅读第 2 章，请先花几分钟阅读本章。

1.1 CoffeeScript 的 REPL

CoffeeScript 自带了功能强大的 REPL[①]（Read-eval-print loop）——一个可交互的控制台，准备好了的话，你可以用它立即开始使用 CoffeeScript。

使用 REPL 非常简单。在你喜欢的终端窗口输入如下这条简单命令即可：

```
> coffee
```

随后，应当能看到如下提示：

```
coffee>
```

如果看到了 `coffee` 的提示，我们就可以开始 CoffeeScript 之旅了。

下面是另一种启动 REPL 的方式：

① http://en.wikipedia.org/wiki/Read-eval-print_loop

```
> coffee -i
```

它是下面这种方式的缩写：

```
> coffee --interactive
```

让我们先看一个简单的例子。在控制台输入如下命令：

```
coffee> 2 + 2
```

控制台应该会显示如下答案：

```
coffee> 4
```

恭喜！你成功迈出了编写 CoffeeScript 代码的第一步。

好，接下来我们来点更有趣的——更具 CoffeeScript“味道”的。不用太关心下面这段代码究竟做了什么（后续我们会做详细解释），只管运行它。

例（源代码：repl1.coffee）

```
a = [1..10]
b = (x * x for x in a)
console.log b
```

输出（源代码：repl1.coffee）

```
[ 1, 4, 9, 16, 25, 36, 49, 64, 81, 100 ]
```

上述这段 CoffeeScript 脚本优美多了吧？简单来说，我们创建了一个新数组，并填充了从 1 到 10 的 10 个数字。然后，循环数组中的这些数字，将它们乘以自身的值并创建第二个数组 b，b 中包含了那些新的值。神奇吧？我说过，后续我会很乐意解释上述代码的具体工作原理的。现在，让我们先享受一下可以不用写那么多 JavaScript 带来的喜悦。不过，你要是对上述代码编译后的 JavaScript 代码是什么样子很好奇，下面就是编译后的 JavaScript 代码。

例（源代码：repl1.js）

```
(function() {
  var a, b, x;

  a = [1, 2, 3, 4, 5, 6, 7, 8, 9, 10];

  b = (function() {
    var _i, _len, _results;
    _results = [];
    for (_i = 0, _len = a.length; _i < _len; _i++) {
      x = a[_i];
```

```
    _results.push(x * x);
   }
   return _results;
  })();

  console.log(b);

}).call(this);
```

如你所见，REPL 是一个很有趣的尝试，可以试验不同的想法。然而，CoffeeScript REPL 并非尽善尽美。你将会看到，CoffeeScript 中设计的空格是有意义的。当我们在 REPL 中尝试书写多行 CoffeeScript 的时候，这种设计就带来了一点小麻烦。解决方案就是使用\字符。

我们来试着写一个简单的 add 函数，函数接收两个数字，返回这两个数的和。在 REPL 中输入如下代码。

例 （源代码：repl2.coffee）

```
add = (x, y)->\
  x + y
console.log add(1, 2)
```

输出 （源代码：repl2.coffee）

```
3
```

注意，我们在第一行代码末尾加了\。这就等于告诉 REPL，我们想要在该表达式中添加多行代码。要记住，如果想要将一行代码添加到表达式中，就要在行末添加\，这一点很重要。首行不以\结束的代码就会被认为是表达式的结束，REPL 会执行该表达式。

还有一点也很重要，要注意\之后还要进行缩进，这样 CoffeeScript 才可以正确地解析这行表达式并将其放在正确的位置。

最后，要退出 REPL，只要按下 Ctrl + C 进程就会结束。

REPL 是一个功能强大的工具，同时也为尝试一些想法提供了便捷的途径，但是，正如我们所看到的，当处理一些复杂代码的时候，它就显得力不从心了。在本章后续的 1.5.5 节“执行 CoffeeScript 文件”部分，我们会讨论如何执行一个包含 CoffeeScript 代码的文件，这是一种比较好的执行复杂代码的方式。

1.2 浏览器端编译

在开发 Web 应用时，有时会想要在 HTML[1]文件中直接内联 CoffeeScript 脚本。CoffeeScript 支持这种方式，后续我会介绍如何来实现。不过，我要先给你提个醒—— 最好不要这么做。首

① http://en.wikipedia.org/wiki/Html

先，最近诸如像非侵入式 JavaScript（Unobtrusive JavaScript[①]）这样的实践非常流行是有其道理的。没错，能在浏览器端执行 CoffeeScript 的确是挺酷的，不过，这真的不是最好的编译方式。将 JavaScript 从 HTML 层分离出来，放在几个单独的文件中，可以保持代码的整洁，并且在不支持 JavaScript 的环境中可以更方便地进行优雅降级。

刚开始书写非侵入式 JavaScript 时会有点困惑，感觉比较难，不过，一段时间之后，写出重用性好、逻辑性强的代码就会显得容易些。使用如 jQuery 这样的工具，你可以等待页面载入后，用 JavaScript 来操作页面上的对象。不过，有的时候，还得要自己亲力亲为去手动实现。通常，都是调用一个 init 方法，可能还会传入 JSON[②]参数来达到这一目的。我鼓励你用纯 JavaScript 来写这段代码。不过，如果你真的很想要用 CoffeeScript 来实现，CoffeeScript 也提供了这样一种在浏览器端编译的方式。

让我们来看一个内嵌了一小段 CoffeeScript 脚本的 HTML 文件。

例（源代码：hello_world.html）

```
<html>
  <head>
    <title>Hello World</title>
    <script src='http://jashkenas.github.com/coffee-script/extras/coffee-script.js'
➥type='text/javascript'&gt;&lt;/script>
  </head>
  <body>
    <script type='text/coffeescript'>
      name = prompt "What is your name?"
      alert "Hello, #{name}"
    </script>
  </body>
</html>
```

因为浏览器原生不支持 CoffeeScript 的编译，至少截止到本书撰写时还不支持，所以，需要在页面中引入 CoffeeScript 的编译器。幸运的是，CoffeeScript 团队考虑到了这点，提供了一个可嵌入的编译器。将如下代码添加到 HTML 文件的 head 部分，就可以嵌入一个 CoffeeScript 编译器：

```
<script src='http://jashkenas.github.com/coffee-script/extras/coffee-script.js'
type='text/javascript'></script>
```

当然，如果你愿意的话，也可以将 coffee-script.js 文件下载下来，存储到本地的项目中。

此外就只需要确保设置了正确的 script 标签类型，让编译器能够获取到内联的 CoffeeScript 脚本，如下所示：

```
<script type='text/coffeescript'></script>
```

在页面载入的时候，coffee-script.js 中的 CoffeeScript 编译器会搜索 HTML 文档，找到类型为 text/coffeescript 的 script 标签，读取其中的内容，将其编译成等效的

① http://en.wikipedia.org/wiki/Unobtrusive_JavaScript
② http://en.wikipedia.org/wiki/Json

JavaScript 代码，然后执行编译后的代码。

1.3　警告

现在你已经知道了如何在 HTML 文档中编译内联的 CoffeeScript 脚本，接下来我要指出以下几点。首先，本书中讨论的所有与作用域、匿名函数包装器等相关的内容，在编译 CoffeeScript 的时候也同样适用。因此，在书写类似代码的时候，要将这一点铭记于心，这非常重要。

其次，或许也是最有用的一点，就是这并不是编译 CoffeeScript 脚本最快的方式。在应用中部署该代码就意味着所有使用它的用户都需要下载这额外的 162.26 KB 的文件来编译 CoffeeScript 脚本。随后，在页面载入后，编译器还得在页面中搜索 `text/coffeescript` 标签，编译其中的代码，然后才能执行。这种用户体验并不好。

知道了这两点后，我希望在你部署前，能够选择正确的方式——进行离线编译 CoffeeScript 脚本。

1.4　命令行编译

尽管在浏览器中执行 CoffeeScript 脚本很有用也相当容易，但这真的不是最好的编译 Coffee Script 的方式。我们应当在部署前就编译好 CoffeeScript 脚本。不过，完全有可能你写的 Node 应用或者其他服务器端的应用压根不在浏览器端，那么浏览器端编译 CoffeeScript 就没什么用了。

那么，怎么样才算是最好的编译 CoffeeScript 脚本的方式呢？好问题。你可以找到很多第三方用于编译 CoffeeScript 脚本的库（不同的平台不同的语言下都有），但是，重要的是，你要理解如何编译 CoffeeScript 脚本，如果有必要，你甚至可以自己写编译脚本。

`compile` 标志

让我们从 `coffee` 命令中最重要的标志 `-c` 开始。`-c` 标志会读取你传递给它的 CoffeeScript 脚本，并在同样的位置编译出 JavaScript 文件。这也是本书中例子的编译方式。

我们继续来创建一个 `hello_world.coffee` 的简单文件，其内容如下：

```
greeting = "Hello, World!"
console.log greeting
```

现在，我们输入如下命令来编译该文件：

```
> coffee -c hello_world.coffee
```

此命令会在 `coffee` 文件所在的目录下，将 CoffeeScript 脚本编译成名为 `hello_world.js` 的新的 JavaScript 文件，该文件的内容如下：

```
(function() {
  var greeting;

  greeting = "Hello, World!";

  console.log(greeting);

}).call(this);
```

好了，这样 `hello_world.js` 文件就可以部署了！该是喝杯咖啡休息一会儿的时候了。

1.5　CoffeeScript 命令行界面

我们已经“玩过”REPL，并学会了如何使用命令行 `coffee` 工具来编译 CoffeeScript 脚本，而 `coffee` 命令还提供了其他一些有趣的选项，值得一看。在终端输入如下命令可以查看所有 `coffee` 命令提供的选项列表：

```
> coffee --help
```

下面是输出结果：

```
Usage: coffee [options] path/to/script.coffee

  -c, --compile       compile to JavaScript and save as .js files
  -i, --interactive   run an interactive CoffeeScript REPL
  -o, --output        set the directory for compiled JavaScript
  -j, --join          concatenate the scripts before compiling
  -w, --watch         watch scripts for changes, and recompile
  -p, --print         print the compiled JavaScript to stdout
  -l, --lint          pipe the compiled JavaScript through JavaScript Lint
  -s, --stdio         listen for and compile scripts over stdio
  -e, --eval          compile a string from the command line
  -r, --require       require a library before executing your script
  -b, --bare          compile without the top-level function wrapper
  -t, --tokens        print the tokens that the lexer produces
  -n, --nodes         print the parse tree that Jison produces
      --nodejs        pass options through to the "node" binary
  -v, --version       display CoffeeScript version
  -h, --help          display this help message
```

现在，我们来进一步介绍其中一些选项。

1.5.1　`output` 标志

当只是试验性地尝试 CoffeeScript 时，完全可以将编译后的 JavaScript 文件和 CoffeeScript 文件放在同一目录下，不过，你可能想要将所有编译后的 JavaScript 文件都放在单独的目录中。那么，如何把我们的 hello_world.coffee 文件编译到比方说 public/javascripts 目录中呢？

答案很简单：在终端命令中加入-o 标志就可以了。

```
> coffee -o public/javascripts -c hello_world.coffee
```

此时查看 public/javascripts 目录，就会发现编译后的 hello_world.js 文件已经在那儿了。

-o 标志不允许更改文件名，所以编译后的 JavaScript 文件和原 CoffeeScript 文件的文件名相同，只是两者后缀不同，前者是.js 而后者是.coffee。因为无法用 coffee 命令来更改文件名，你或许会想更改源文件名，或者写一段脚本来完成这样的任务：编译 CoffeeScript 文件，然后将其文件名改为你喜欢的名字。

1.5.2　bare 标志

在本书后续部分会看到，CoffeeScript 编译时，会将编译后的 JavaScript 代码包在一个匿名函数中。这部分内容会在第 2 章中详细介绍，所以这里先不深究。下面就是一个将生成的 JavaScript 代码包在匿名函数中的例子。

例（源代码：hello_world.js）

```
(function() {
  var greeting;

  greeting = "Hello, World!";

  console.log(greeting);

}).call(this);
```

有时，出于某种原因，可能并不想要这个匿名函数包装器。在这种时候，可以给 CoffeeScript 编译器传递-b 标志。

```
> coffee -b -c hello_world.coffee
```

这样将会把我们的 CoffeeScript 脚本编译为如下 JavaScript 代码。

例（源代码：hello_world_bare.js）

```
var greeting;
greeting = "Hello, World!";
console.log(greeting);
```

现在知道如何移除这个匿名函数包装器了吧，不过我要提醒你的是，之所以默认编译器会将 JavaScript 代码包在匿名函数中，是有其重要原因的。如果想要了解更多与此匿名函数相关的内容，请看第2章。

1.5.3 print 标志

有时，在编译 CoffeeScript 文件时，会想要直接看到编译后的内容。幸运的是，coffee 命令提供了-p 标志：

```
> coffee -p hello_world.coffee
```

此命令会将结果直接输出到终端，如下所示：

```
(function() {
  var greeting;

  greeting = "Hello, World!";

  console.log(greeting);

}).call(this);
```

这对调试来说无疑是非常有用的，或者可以作为很好的学习 CoffeeScript 的工具。通过比较 CoffeeScript 和编译后的 JavaScript（如我们在本书中做的那样），你会渐渐明白 CoffeeScript 内部究竟做了什么。我刚学 CoffeeScript 的时候，它帮了我很大的忙。通过用它进一步了解编译器的处理方式，也帮助我成为了一名更好的 JavaScript 开发者。

1.5.4 watch 标志

在开发 CoffeeScript 项目时，可能有人不愿意老在命令行进行编译。CoffeeScript 也考虑到了这点，它提供了-w 参数，可以让编译变得更加容易。有了这个参数，就等于告诉 coffee 命令始终监听 CoffeeScript 文件，一旦文件发生变化就立即重新编译。下面是一个例子：

```
> coffee -w -c app/assets/coffeescript
```

键入上述命令之后，app/assets/coffeescript 目录及其子目录所有.coffee 后缀的文件在任何时候被“触碰”（touched[①]）发生任何变化，都会自动被重新编译。

在 CoffeeScript 1.2.0 中，-w 参数会监听所有在被监听目录中新增的文件。不过，以我的经验看，这可能是因为底层 Node 的问题导致的 bug。希望，在你阅读本书的时候，这些问题已经

① 触碰一个文件在不同的操作系统中代表了不同的操作，不过，通常，保存文件就属于“触碰”，并足以触发-w 去进行重新编译。

得以解决了。不过，不用担心，还有大量的第三方工具可以用来监听文件系统的相关事件，如：文件添加与移除。截止到本书撰写时，我个人倾向于使用 Guard[①]。它是一个 Ruby gem，实现监听这类事件并执行一些自定义代码，比方说，当这些事件触发的时候，编译 CoffeeScript 文件。

> **提示**：除了 Guard 之外，还可以试试 Trevor Burnham 的 Jitter，它实现了与 Guard 类似的目标：监听并编译所有的 CoffeeScript 文件。它也是用 CoffeeScript 写的，值得一试。

1.5.5　执行 CoffeeScript 文件

我们已经介绍了编译 CoffeeScript 的几种方式，同时也讨论了在编译 CoffeeScript 时传递给 `coffee` 命令的一些选项，但是，怎么样执行 CoffeeScript 文件呢？你可能会用 CoffeeScript 写一个 Web 服务器，甚至只是写一个处理简单数字运算的脚本。然后，可以使用你已经学到的工具进行编译，接着将它们外链到 HTML 文件中，并最终在浏览器中执行。这种方式适用于简单脚本，而对于一些诸如 Web 服务器这样复杂的情况则不然，并且也不具实践意义。

为了解决这类问题，`coffee` 命令允许直接执行 CoffeeScript 文件，如下所示：

```
> coffee hello_world.coffee
```

本书中大部分的例子也都可以使用这种方式来运行（特殊说明的除外）。

1.5.6　其他选项

还有其他一些诸如`-n`、`-t` 这样的选项。尽管这些选项能够给你一些非常有趣的输出结果，让你看到 CoffeeScript 编译时是如何处理的，不过，它们在本书中并不会给我们带来太大帮助，所以，这里我不做介绍。但是，我还是鼓励你花些时间去试试这些选项，看看它们会产生什么。通过阅读在线的 CoffeeScript 命令的带注解的源代码[②]，可以了解更多关于这些选项的信息。

1.6　小结

本章介绍了几种不同的编译和执行 CoffeeScript 代码的方式。同时也介绍了这些编译 CoffeeScript 方式的优缺点，现在我们已经掌握了理解本书后续例子所需的知识。最后，我们深入探讨了 `coffee` 命令，学习了可以传递给它的最重要的一些选项和参数。

---

① https://github.com/guard/guard
② http://jashkenas.github.com/coffee-script/documentation/docs/command.html

第2章 基础知识

我们已经介绍了编译和执行 CoffeeScript 这种比较枯燥的内容，下面我们开始介绍如何书写 CoffeeScript。

本章将介绍 CoffeeScript 的语法：标点符号、作用域、变量以及其他一些小东西。

2.1 语法

对 CoffeeScript 最大的争议莫过于它的语法了，特别是它没有标点符号这一点。诸如大括号、分号这样的标点符号在 CoffeeScript 的世界中是绝种的，连括号都是濒临灭绝的"物种"。

为了证明这一点，我们来看一段你或许非常熟悉的 JavaScript 代码。下面这段 jQuery 代码，用于发送一个远程 AJAX 请求，然后对返回的结果做相应的处理。

例 （源代码：jquery_example.js）

```
$(function() {
  $.get('example.php', function(data) {
    if (data.errors != null) {
      alert("There was an error!");
    } else {
      $("#content").text(data.message);
    }
  }, 'json')
})
```

针对上述例子，CoffeeScript 允许我们忽略大量额外的标点符号。下面是用 CoffeeScript 写的等效代码。

例 （源代码：jquery_as_coffee.coffee）

```
$ ->
  $.get 'example.php', (data) ->
      if data.errors?
        alert "There was an error!"
      else
        $("#content").text data.message
    , 'json'
```

本书后续部分会详细介绍上述例子的细节，不过，现在先来对比一下上述两段代码，看看我们的 CoffeeScript 代码中移除了哪些 JavaScript 中的内容。

2.1.1 有意义的空格

首先，我们移除了所有的大括号和分号。

例 （源代码：jquery.js）

```
$(function()
  $.get('example.php', function(data)
    if (data.errors != null)
      alert("There was an error!")
    else
      $("#content").text(data.message)
  , 'json')
)
```

移除了之后是怎么工作的呢？CoffeeScript 是怎样正确解析这段代码的呢？答案很简单，奥妙就是我们几乎天天要写的：空格！和 Python 一样，CoffeeScript 利用有意义的空格来解析表达式。

听说，有些不喜欢这种有意义的空格的人对其充满抱怨。我发现这种抱怨多少有点儿不同寻常。难道你会像下面这样来书写上述 JavaScript 代码吗？

例 （源代码：jquery.js）

```
$(function() {
$.get('example.php', function(data) {
if (data.errors != null) {
alert("There was an error!");
} else {
$("#content").text(data.message);
}
}, 'json')
})
```

我希望你不会！如果你真的用这种方式书写 JavaScript，那我请求你帮帮你的同伴（和你一样用这种方式书写 JavaScript 的开发者），多花些时间确保你们的代码是正确缩进的。可读性是可维护性的关键，同时也是帮助你将 JavaScript 代码转化为 CoffeeScript 的关键。

使用有意义的空格，当你将 if 语句下一行代码进行缩进时，CoffeeScript 编译器就能知道这行代码属于 if 代码块。编译器下次遇到和 if 语句同样级别的缩进的时候，它就能知道该 if 语句已经结束并将该代码视作与 if 同级别去执行。

下面是一个简单的例子，展示了在没有正确格式化 CoffeeScript 时会出现的错误类型。

例（源代码：whitespace.coffee）

```
for num in [1..3]
     if num is 1
       console.log num
        console.log num * 2
  if num is 2
         console.log num
         console.log num * 2
```

输出（源代码：whitespace.coffee）

```
Error: In content/the_basics/whitespace.coffee, Parse error on line 4: Unexpected
➥'INDENT'
     at Object.parseError (/usr/local/lib/node_modules/coffee-script/lib/coffee-script/
➥parser.js:470:11)
     at Object.parse (/usr/local/lib/node_modules/coffee-script/lib/coffee-script/
➥parser.js:546:22)
    at /usr/local/lib/node_modules/coffee-script/lib/coffee-script/coffee-script.
➥js:40:22
     at Object.run (/usr/local/lib/node_modules/coffee-script/lib/coffee-script/
➥coffee-script.js:68:34)
     at /usr/local/lib/node_modules/coffee-script/lib/coffee-script/command.js:135:29
     at /usr/local/lib/node_modules/coffee-script/lib/coffee-script/command.js:110:18
     at [object Object].<anonymous> (fs.js:114:5)
     at [object Object].emit (events.js:64:17)
     at afterRead (fs.js:1081:12)
     at Object.wrapper [as oncomplete] (fs.js:252:17)
```

2.1.2 function 关键字

接下来我们把 function 关键字给移除了。写 CoffeeScript 代码不再需要写 function 关键字了。

例（源代码：jquery.js）

```
$( ()->
  $.get('example.php', (data)->
    if (data.errors != null)
      alert("There was an error!")
    else
```

```
    $("#content").text(data.message)
  , 'json')
)
```

我们可以在函数参数列表的右边用->箭头来替换 function 关键字。我知道，一开始会有些难记而且不容易理解，不过，试着想想代码流以及参数的位置，就会觉得这种箭头的方式事实上很形象。

2.1.3 括号

再接着，我们移除了示例代码中的大部分括号，但并非全部。

例（源代码：jquery.js）

```
$ ->
  $.get 'example.php', (data)->
    if data.errors != null
      alert "There was an error!"
    else
      $("#content").text data.message
  , 'json'
```

为何我们不将所有的括号都移除呢？绝大部分情况下，移除括号是可选的。何时该使用括号的规则有点混乱，特别是涉及到函数的时候。现在来看看我们刚写的那段代码，看看为何有些括号被保留了，有些却被删除了。

当调用 alert 函数时，如下所示：

```
alert "There was an error!"
```

这里可以将括号移除，原因是当调用一个函数同时还传递参数时，括号可以省略。然而，当调用函数又无须传递参数时，就需要加上括号，为的是让 JavaScript 知道这是调用函数，而不是访问变量。正如我之前所说：有点儿混乱。

> **提示**：当对是否移除括号举棋不定的时候，若觉得加上括号可以让代码更整洁、更有可读性，就可以加上括号。

既然在调用函数并传递参数时，可以省略括号，那为何我们还要将代码写成如下形式呢？

```
$("#content").text data.message
```

为什么不直接写成下面这样呢？

```
$ "#content" .text data.message
```

如果这样写的话，上述代码编译后的 JavaScript 代码会是如下形式：

```
$("#content".text(data.message));
```

正如你所见，CoffeeScript 不确定你要调用的是哪个对象上的 text 函数，所以，它假定是字符串"#content"上的 text 函数。这里我们保留括号，就是为了明确告诉 CoffeeScript，调用的是$("#content")返回的 jQuery 对象上的 text 方法。

在结束括号的相关讨论之前（别担心，在介绍函数的时候还会再次提到括号的），我想要指出的是，括号还可以用作逻辑分组运算。

例（源代码：grouping.coffee）

```
if x is true and (y is true or z is true)
  console.log 'hello, world'
```

例（源代码：grouping.js）

```
(function() {

  if (x === true && (y === true || z === true)) console.log('hello, world');

}).call(this);
```

2.2 作用域与变量

本节内容中，我们会介绍 CoffeeScript 中作用域以及变量是如何定义和工作的。在 JavaScript 中，这是个棘手的问题，通常也是 bug 和困惑的来源。本节中，我们会看到 CoffeeScript 是如何让 bug 与作用域间的关系成为过去式的。

2.2.1 JavaScript 中的变量作用域

很多人，包括新手和有一定经验的开发者，都不知道在 JavaScript 中有两种声明变量的方式。其中或许有人知道这两种方式，但却未必知道这两者的区别。正因如此，让我们简单来看下这两种方式，至少从基础层面理解这两种方式的区别。

来看下下面这段代码。

例（源代码：js_variable_scope.js）

```
a = 'A';
myFunc = function() {
  a = 'AAA';
  var b = 'B';
}
```

如果在浏览器或者其他的 JavaScript 引擎中运行上述代码，我们会得到如下结果。

输出：

```
> console.log(a)
A
> myFunc();
> console.log(a)
AAA
> console.log(b)
ReferenceError: b is not defined
```

当尝试访问变量 b 时，结果抛出了错误，这或许是你预期的。但是，你可能意想不到的是，在 myFunc 函数中对变量 a 的赋值居然会影响到 a 的值，对吗？那究竟为什么会这样呢？

答案很简单，其根本原因就是这两个变量的定义方式不同。

当不使用 var 关键字定义变量时，就等于告诉 JavaScript 创建一个全局的变量。因为变量 a 已经在全局作用域中存在了，所以，很遗憾，在 myFunc 中对 a 定义的新值就替换了原先的值。

变量 b 是在 myFunc 函数内部使用 var 关键字进行定义的，这就等于告诉 JavaScript 创建一个名为 b 的变量，并且其作用域只在 myFunc 函数中。正因为变量 b 的作用域在函数内部，所以当我们在函数外部访问 b 时，就会抛出错误，因为这个时候，JavaScript 在全局作用域中根本找不到 b 了。

> **提示：**本节的例子就证明了为什么在定义变量的时候都应该使用 var 关键字。这绝对是最佳实践，并且 CoffeeScript 会帮助你始终使用 var 来定义变量。

2.2.2　CoffeeScript 中的变量作用域

现在，我们已经对 JavaScript 中的变量作用域有所了解，下面让我们再来看一下此前的例子，不同的是，这次用 CoffeeScript 来写。

例（源代码：coffeescript_variable_scope.coffee）

```
a = 'A'
myFunc = ->
  a = 'AAA'
  b = 'B'
```

例（源代码：coffeescript_variable_scope.js）

```
(function() {
  var a, myFunc;

  a = 'A';

  myFunc = function() {
```

```
    var b;
    a = 'AAA';
    return b = 'B';
  };

}).call(this);
```

暂时先忽略上述代码最外层的匿名封装器函数（后续很快会对其做详细介绍），让我们先来看看 CoffeeScript 是如何处理变量的声明的。我们注意到，每一个变量，包括指向 myFunc 函数的变量，都会用 var 关键字来声明。CoffeeScript 会确保使用正确的方式来声明变量。

上述代码中还有另外很有意思的一点，尽管 CoffeeScript 帮助我们进行正确的变量声明，但它并未在 myFunc 函数中对变量 a 再次进行声明。这是因为，当 CoffeeScript 进行编译时，它看到之前已经定义了变量 a，它假定你会在 myFunc 函数中继续使用该变量。

2.2.3 匿名封装器函数

你看到并且也应该注意到了，所有编译后的 JavaScript 代码都会被封装到一个匿名的自运行的函数中。想必现在你对这个函数到底是做什么用的肯定充满好奇。那就让我来告诉你吧。

正如我们此前在 JavaScript 的变量作用域的例子中看到的，对于不用 var 关键字定义的变量，可以很容易地访问到。这样声明的变量会始终属于全局作用域，可以被很容易地访问到。

即使我们定义 a 变量时使用了 var 关键字，它仍属于全局作用域。你可能会问，为什么会这样呢？原因很简单，我们在函数外定义了变量 a。因为该变量是在函数作用域外定义的，所以它在全局作用域都可用。这意味着，如果有另外一个你使用的库中也定义了变量 a，那么后面定义的 a 会把前面定义的 a 覆盖掉。当然了，在所有的作用域中都是如此，并非仅仅是全局作用域中。

那么如何在全局作用域中定义变量、函数，才能只让程序本身可以访问而别人无法访问呢？使用匿名封装器函数将代码封装起来。这也正是 CoffeeScript 将代码编译为 JavaScript 代码时的做法。

现在，看到这里你可能会想到两件事情。第一，“使用匿名封装器函数这种做法的确很聪明，使得可以在其内部随意操作也不会污染全局作用域。”第二，你应当问问自己，“等等，那如何才能将变量、函数暴露到全局作用域中，供外部访问呢？”这两点非常重要。我们来解决第二个问题，因为这是一种很常见的需求。

如果你将整个程序或者库都写在匿名函数内部的话，那么就和 CoffeeScript 强制的一样了，其他的库或代码都无法访问你的代码。有的时候，我们的确希望如此，但是如果你并不希望如此的话，这里也有几种解决方案。

下面这个例子展示了将一个函数共享到外部的方法。

例 通过 window 对象共享函数（源代码：expose_with_window.coffee）

```
window.sayHi = ->
  console.log "Hello, World!"
```

例 通过 window 对象共享函数（源代码：expose_with_window.js）

```
(function() {

  window.sayHi = function() {
    return console.log("Hello, World!");
  };

}).call(this);
```

上述例子中，使用了 window 对象来对外共享 sayHi 函数。如果代码在浏览器端运行，这就是一种非常好的对外共享函数的方式。然而，随着 Node.JS 以及其他服务器端 JavaScript 技术的崛起，越来越流行在这类环境而非浏览器环境中运行 JavaScript 代码。如果我们直接使用 coffee 命令来执行这段代码的话，会得到如下输出结果。

例 通过 window 对象共享函数（源代码：expose_with_window.coffee.output）

```
ReferenceError: window is not defined
    at Object.<anonymous> (.../the_basics/expose_with_window.coffee:3:3)
    at Object.<anonymous> (.../the_basics/expose_with_window.coffee:7:4)
    at Module._compile (module.js:432:26)
    at Object.run  (/usr/local/lib/node_modules/coffee-script/lib/coffee-script/
➥coffee-script.js:68:25)
    at /usr/local/lib/node_modules/coffee-script/lib/coffee-script/command.js:135:29
    at /usr/local/lib/node_modules/coffee-script/lib/coffee-script/command.js:110:18
    at [object Object].<anonymous> (fs.js:114:5)
    at [object Object].emit (events.js:64:17)
    at afterRead (fs.js:1081:12)
    at Object.wrapper [as oncomplete] (fs.js:252:17)
```

在用 coffee 命令行运行时，window 对象是不存在的，自然也无法通过它来共享函数。那么如何解决这个问题呢？答案也很简单。在 CoffeeScript 为我们创建匿名函数封装器的同时，也将 this 对象传递进来了。因此，只要将函数绑定到 this 对象上就可以实现函数的共享了。看下面的例子。

例 通过 this 共享函数（源代码：expose_with_this.coffee）

```
this.sayHi = ->
  console.log "Hello, World!"
```

例 通过 this 共享函数（源代码：expose_with_this.js）

```
(function() {

  this.sayHi = function() {
    return console.log("Hello, World!");
  };

}).call(this);
```

当在浏览器端执行上述代码时，sayHi 函数会共享给外部的 JavaScript 代码，因为浏览器中的 this 就指向 window 对象。当用 coffee 命令或者在 Node.JS 环境中执行时，this 就会指向全局对象。使用 this 来共享变量和函数，既能提高代码健壮性又能获得平台保障性。

出于完整性的考虑，下面给出一个更简单、更整洁的共享代码的方式。

例 通过@共享函数（源代码：expose_with_at.coffee）

```
@sayHi = ->
  console.log "Hello, World!"
```

例 通过@共享函数（源代码：expose_with_at.js）

```
(function() {

  this.sayHi = function() {
    return console.log("Hello, World!");
  };

}).call(this);
```

正如在编译后 JavaScript 代码中看到的那样，在 CoffeeScript 中，@标志会被编译为 this。在 CoffeeScript 中，任何你想使用 this 的地方，都可以使用@来替代，效果是一样的。尽管使用@并非强制的，不过在 CoffeeScript 中推荐使用它，在本书中，我也都会使用@。

2.3 插值

本节将介绍 CoffeeScript 中的字符串插值、heredoc 以及注释。字符串插值可以让我们轻松地构建动态字符串，而不必担心令人困惑、易错的合并语法。heredoc 可以让我们很容易地构建格式良好的多行字符串。注释部分就不言自明了。

2.3.1 字符串插值

JavaScript 中，最令我懊恼的事情之一就是构建动态字符串。来看下面这个例子，用一些动态属性在 JavaScript 中构建一个 HTML 文本域。

例 （源代码：javascript_concatenation.js）

```
var field, someId, someName, someValue;
someName = 'user[firstName]';
someId = 'firstName';
someValue = 'Bob Example';
field = "<input type='text' name='" + someName + "' id='" + someId + "' value='" +
➥(escape(someValue)) + "'>";
console.log(field);
```

输出（源代码：javascript_concatenation.js）

```
<input type='text' name='user[firstName]' id='firstName' value='Bob%20Example'>
```

看到了吧，多么丑陋、令人困惑而且暗藏 bug 的代码啊！我有没有正确地结束单引号（'）？双引号（"）的数量对吗？我想应该没问题，不过，这确实不是一眼就能够看清楚的。

CoffeeScript 沿用了诸如 Ruby 这样的更现代的语言的概念，提供了两种不同类型的字符串：插值字符串和文本字符串。我们来看看。

2.3.2　插值字符串

为了摆脱我们在 HTML 文本域的例子中看到的令人讨厌的嵌套拼接字符串，CoffeeScript 允许使用字符串插值来解决这个问题。

什么是字符串插值？字符串插值是这样一种方式：允许在字符串中注入任意的 CoffeeScript 代码，并在运行时执行。在我们的例子中，我们想要将一些变量注入到 HTML 字符串中，CoffeeScript 允许我们这样做。下面是一个等效的例子，只是这次使用 CoffeeScript 的字符串插值来实现。

例（源代码：html_string_interpolation.coffee）

```
someName = 'user[firstName]'
someId = 'firstName'
someValue = 'Bob Example'

field = "<input type='text' name='#{someName}' id='#{someId}' value='#{escape
➥someValue}'>"

console.log field
```

例（源代码：html_string_interpolation.js）

```
(function() {
  var field, someId, someName, someValue;

  someName = 'user[firstName]';

  someId = 'firstName';

  someValue = 'Bob Example';

  field = "<input type='text' name='" + someName + "' id='" + someId + "' value='" +
➥(escape(someValue)) + "'>";

  console.log(field);

}).call(this);
```

输出（源代码：html_string_interpolation.coffee）

```
<input type='text' name='user[firstName]' id='firstName' value='Bob%20Example'>
```

代码是不是更好看了？这样的代码，更具可读性和可维护性，并且书写也很方便。

如你所知，在 JavaScript 中，并没有插值字符串这类东西。所有的字符串都是一样的。而在 CoffeeScript 中则不然，不同类型字符串处理方式是不同的。比方说，我们刚刚所使用的，用双引号括起来的字符串，如果有必要的话，CoffeeScript 编译器会将该字符串转化为拼接的 JavaScript 字符串。而用单引号括起来的字符串在 CoffeeScript 中被称为文本字符串，稍后就会对其作相应的介绍。

要在双引号字符串中注入 CoffeeScript 脚本，可以使用#{}语法。所有在这对大括号中的代码都会被编译器单独解析，并将其结果拼接到该字符串上。大括号中的代码可以是任意有效的 CoffeeScript 代码。

例（源代码：string_interpolation_extra.coffee）

```
text = "Add numbers: #{1 + 1}"
console.log text

text = "Call a function: #{escape "Hello, World!"}"
console.log text

day = 'Sunday'
console.log "It's a beautiful #{if day is 'Sunday' then day else "Day"}"
```

例（源代码：string_interpolation_extra.js）

```
(function() {
  var day, text;

  text = "Add numbers: " + (1 + 1);

  console.log(text);

  text = "Call a function: " + (escape("Hello, World!"));

  console.log(text);

  day = 'Sunday';

  console.log("It's a beautiful " + (day === 'Sunday' ? day : "Day"));

}).call(this);
```

输出（源代码：string_interpolation_extra.coffee）

```
Add numbers: 2
Call a function: Hello%2C%20World%21
It's a beautiful Sunday
```

2.3.3 文本字符串

顾名思义，文本字符串就是：字符串中的内容是什么，最后执行输出的就是什么。JavaScript中就只有文本字符串。

在 CoffeeScript 中使用单引号（'）来构建文本字符串。让我们再来看看此前构造的 HTML 文本域的例子。这次，我们使用单引号而不是双引号来构建，看看会发生什么。

例 （源代码：html_string_literal.coffee）

```
someName = 'user[firstName]'
someId = 'firstName'
someValue = 'Bob Example'

field = '<input type=\'text\' name=\'#{someName}\'
➥id=\'#{someId}\' value=\'#{escape(someValue)}\'>'

console.log field
```

例 （源代码：html_string_literal.js）

```
(function() {
  var field, someId, someName, someValue;

  someName = 'user[firstName]';

  someId = 'firstName';

  someValue = 'Bob Example';

  field = '<input type=\'text\' name=\'#{someName}\'
➥id=\'#{someId}\' value=\'#{escape(someValue)}\'>';

  console.log(field);

}).call(this);
```

输出 （源代码：html_string_literal.coffee）

```
<input type='text' name='#{someName}' id='#{someId}'value='#{escape(someValue)}'>
```

从输出可以看出，我们并没有获得预期的输出结果。这是因为文本字符串并不支持字符串插值。最终我们看到的仅仅是动态内容的占位符，而非真正要注入到字符串中的动态内容。有时，我们想要的就是这种效果，但这种情况相对比较少见。

尽管文本字符串不允许注入动态内容，但和 JavaScript 一样，它还支持少量的运算和解析。CoffeeScript 中，还允许在文本字符串中使用常用的转义字符。如下所示。

例（源代码：literal_string_with_escapes.coffee）

```
text = 'Header\n\tIndented Text'
console.log text
```

例（源代码：literal_string_with_escapes.js）

```
(function() {
  var text;

  text = 'Header\n\tIndented Text';

  console.log(text);

}).call(this);
```

输出（源代码：literal_string_with_escapes.coffee）

```
Header
        Indented Text
```

如你所见，换行符（\n）和 tab 符号（\t）都被正确地解释并在输出中被正确地处理了。和 JavaScript 一样，CoffeeScript 允许使用双反斜杠来转义单反斜杠，如下所示。

例（源代码：literal_string_with_backslash.coffee）

```
text = 'Insert \\some\\ slashes!'
console.log text
```

例（源代码：literal_string_with_backslash.js）

```
(function() {
  var text;

  text = 'Insert \\some\\ slashes!';

  console.log(text);

}).call(this);
```

输出（源代码：literal_string_with_backslash.coffee）

```
Insert \some\ slashes!
```

在像 Ruby 这样的语言中，会使用文本字符串来提高性能。这样在程序执行时，运行时环境就不需要去解析字符串，只需要做一些必要的操作。然而，因为 CoffeeScript 最终会被编译为 JavaScript 代码，因此，性能问题也就从运行时转移到了编译时。

> **提示：**使用文本字符串代替插值字符串能提高性能，不过也只提高了编译时的性能。我觉得始终用双引号、插值字符串并没有坏处。即便你现在不用插值字符串，以后要将文本字符串替换为插值字符串也很容易。

2.3.4　heredoc

heredoc[①]也叫 here document，允许在 CoffeeScript 中轻松构建多行字符串，并保留该字符串中所有的空格和换行符。heredoc 字符串采用和插值字符串以及文本字符串同样的规则。要在 CoffeeScript 中构建插值 heredoc 字符串，可以在每个字符串末尾使用三个双引号。要构建文本 heredoc 字符串，则使用三个单引号。

让我们来看一个简单的例子。在之前的 HTML 文本域上再增加一些 HTML。

例（源代码：heredoc.coffee）

```
someName = 'user[firstName]'
someId = 'firstName'
someValue = 'Bob Example'

field = """
        <ul>
          <li>
             <input type='text' name='#{someName}' id='#{someId}'
value='#{escape(someValue)}'>
          </li>
        </ul>
        """

console.log field
```

例（源代码：heredoc.js）

```
(function() {
  var field, someId, someName, someValue;

  someName = 'user[firstName]';

  someId = 'firstName';

  someValue = 'Bob Example';

  field = "<ul>\n <li>\n <input type='text' name='" + someName + "' id='" + someId
➥+ "' value='" + (escape(someValue)) + "'>\n </li>\n</ul>";
```

① http://en.wikipedia.org/wiki/Heredoc

```
  console.log(field);

}).call(this);
```

输出 （源代码：heredoc.coffee）

```
<ul>
  <li>
    <input type='text' name='user[firstName]' id='firstName'value='Bob%20Example'>
  </li>
</ul>
```

如你所见，最终的输出，如原文一样，保留了很好的格式。除此之外，源代码中，heredoc开始部分的缩进也很好地保留了，这使保持代码格式良好变得更加容易。

2.3.5 注释

每一门好的语言都需要多种添加注释的方式，CoffeeScript 也不例外。CoffeeScript 中有两种注释方式，不同的注释方式在最终编译后的 JavaScript 中表现也不同。

2.3.6 内联注释

第一种注释方式是内联注释。内联注释非常简单，只需使用#符号即可创建内联注释。所有在同行的#符号后的内容都会被 CoffeeScript 编译器忽略。

例 （源代码：inline_comment.coffee）

```
# 计算工资
calcPayroll()

payBils() # 支付账单
```

例 （源代码：inline_comment.js）

```
(function() {

  calcPayroll();

  payBils();

}).call(this);
```

上述例子中，可以看到，注释最终并没有出现在 JavaScript 代码中。对于这种方式究竟是好是坏，还有些争论。要是注释会跟到编译后的 JavaScript 代码就再好不过了。不过，从CoffeeScript 直接映射到 JavaScript 并非总那么好，编译器没法总能正确地知道注释该在什么位置。从有利的方面看，正因为没有了注释，才使得编译后的 JavaScript 代码更加轻量级，也正因如此，可以让我们在 CoffeeScript 中随意添加注释，而不用担心会使 JavaScript 代码变得臃肿。

2.3.7 块级注释

CoffeeScript 中另外一种注释是块级注释。块级注释用于多行、内容比较多的注释。这类注释包括许可证、版本信息、API 使用文档，等等。与内联注释不同，CoffeeScript 会将块级注释带到编译后的 JavaScript 代码中。

例 （源代码：block_comment.coffee）

```
###
My Awesome Library v1.0
Copyright: Me!
Released under the MIT License
###
```

例 （源代码：block_comment.js）

```
/*
My Awesome Library v1.0
Copyright: Me!
Released under the MIT License
*/

(function() {


}).call(this);
```

如你所见，定义块级注释和定义 heredoc 类似，通过在注释的两端使用三个#符号就可以定义块级注释了。

2.4 扩展的正则表达式

在如何定义、使用和执行正则表达式[①]方面，CoffeeScript 和 JavaScript 是相同的。而当想要书写那些长的、复杂的正则表达式时，CoffeeScript 则提供了一些帮助。

我们都有这样的经历，有些正则表达式比较难处理，所以我们想要将其拆分成多行，并为每行表达式都添加注释。CoffeeScript 就允许这样做。

与定义 heredoc 和块级注释类似，要定义一个多行正则表达式，只要在表达式的两端使用三个斜杠即可。

① http://en.wikipedia.org/wiki/Regular_expressions

让我们来看一个实际使用多行正则表达式的例子：

例（源代码：extended_regex.coffee）

```
REGEX = /// ^
  (/ (?! [\s=] )   # 不允许开头空格或者等号符号
  [^ [ / \n \\ ]*  # 所有其他内容
  (?:
    (?: \\[\s\S]   # 任何转义的内容
      | \[         # 字符类
          [^ \] \n \\ ]*
          (?: \\[\s\S] [^ \] \n \\ )* )*
        ]
    ) [^ [ / \n \\ ]*
  )*
  /) ([imgy]{0,4}) (?!\w)
///
```

例（源代码：extended_regex.js）

```
(function() {
  var REGEX;

  REGEX = /^(\/(?![\s=])[^[\/\n\\]*(?:(?:\\[\s\S]|\[[^\]\n\\]*(?:\\[\s\S]
➥[^\]\n\\]*)*])[^[\/\n\\]*]*\/)([imgy]{0,4})(?!\w)/;

}).call(this);
```

从编译后的 JavaScript 代码中能看到所有额外的空格和注释都被移除了。

2.5　小结

至此，你已经了解了 CoffeeScript 的语法，我们可以开始介绍 CoffeeScript 中更有趣、更细节的部分。如果你对本章介绍的内容还没搞明白，请再花几分钟回过头去看看。搞懂本章内容是非常重要的，因为这些内容是本书后续章节的基础，所以，在进入下一章节内容前，请确保你已掌握了本章内容。

第3章

控制结构

几乎所有的语言中都有操作符①和条件语句②的概念，JavaScript 和 CoffeeScript 也都不例外。操作符和条件语句“联手”充当了所有编程语言的一个重要角色。操作符允许进行两个数字间的加减法、两个对象间的比较，以及一个对象中的字节移动等操作。条件语句则允许我们根据预先定义的条件来控制应用程序的走向。比如，如果用户未登录，则显示登录界面，否则显示用户界面。这就是条件语句。

本章对操作符和条件语句都会介绍，还会介绍它们在 CoffeeScript 中如何来定义。

3.1 操作符与别名

在操作符方面，JavaScript 和 CoffeeScript 绝大部分的处理都是一致的，并且处理得很好。不过，在有些地方 CoffeeScript 做得更近一步，可以帮助我们避免进入 JavaScript 的陷阱。为了确保我们完全掌握操作符，我们先来看看 JavaScript 中的操作符，其中与 CoffeeScript 不同的地方，我会进行指明。开始前，先申明，我假定你了解 JavaScript 中的操作符。如果你还不了解，那么现在是个好机会。需要这方面资料的话，我推荐 http://en.wikibooks.org/wiki/JavaScript/Operators。文章写得很好，提供了 JavaScript 中操作符的概览。

3.1.1 运算

下面是 JavaScript 中的运算操作符列表：

- + 加

① http://en.wikipedia.org/wiki/Operator_(programming)
② http://en.wikipedia.org/wiki/Conditional_(programming)

- - 减
- * 乘
- / 除（结果返回浮点数）
- % 取模（结果返回余数）
- + 一元运算符，用于将字符串转换为数字
- - 一元运算符，取反
- ++ 递增（可以用作前缀也可以用作后缀）
- -- 递减（可以用作前缀也可以用作后缀）

下面，我们来看下这些操作符是怎样对应到 CoffeeScript 中的。

例（源代码：arithmetic.coffee）

```
console.log "+ 加: #{1 + 1}"

console.log "- 减: #{10 - 1}"

console.log "* 乘: #{5 * 5}"

console.log "/ 除: #{100 / 10}"

console.log "% 取模: #{10 % 3}"

console.log "+ 一元运算，将字符串转换为数字: #{+'100'}"

console.log "- 一元运算，取反: #{-50}"

i= 1
x = ++i
console.log "++ 递增: #{x}"

i= 1
x = --i
console.log "-- 递减: #{x}"
```

例（源代码：arithmetic.js）

```
(function() {
  var i, x;

  console.log("+ 加: " + (1 + 1));

  console.log("- 减: " + (10 - 1));

  console.log("* 乘: " + (5 * 5));

  console.log("/ 除: " + (100 / 10));

  console.log("% 取模: " + (10 % 3));
```

```
  console.log("+ 一元运算，将字符串转换为数字： " + (+'100'));

  console.log("- 一元运算，取反： " + (-50));

  i = 1;

  x = ++i;

  console.log("++ 递增： " + x);

  i = 1;

  x = --i;

  console.log("-- 递减： " + x);

}).call(this);
```

输出 （源代码：arithmetic.coffee）

```
+ 加： 2
- 减： 9
* 乘： 25
/ 除： 10
% 取模： 1
+ 一元运算，将字符串转换为数字： 100
- 一元运算，取反： -50
++ 递增： 2
-- 递减： 0
```

如上述例子所示，所有的 CoffeeScript 运算操作符都直接映射到了 JavaScript 的操作符上，所以不需要花时间去记了。

3.1.2　赋值

我们接着来看一下 JavaScript 中的赋值操作符①，列表如下：

- = 赋值
- += 先加后赋值
- -= 先减后赋值
- *= 先乘后赋值
- /= 先除后赋值
- %= 先取模后赋值
- ?= 判断是否存在，不存在则赋值

① 后三个操作符是 CoffeeScript 引入的，在 JavaScript 中并不存在。——译者注

- ||= 判断条件为假则赋值
- &&= 判断两者都为 true 则赋值

如何对应到 CoffeeScript 中？

例（源代码: assignment.coffee）

```
console.log "= 赋值 ："
x = 10
console.log x

console.log "+= 先加后赋值："
x += 25
console.log x

console.log "-= 先减后赋值 ："
x -= 25
console.log x

console.log "*= 先乘后赋值："
x *= 10
console.log x

console.log "/= 先除后赋值 ："
x /= 10
console.log x

console.log "%= 先取模后赋值 ："
x %= 3
console.log x

console.log "?= 判断是否存在，不存在则赋值 ："
y ?= 3
console.log y
y ?= 100
console.log y

console.log "||= 判断条件为假则赋值："
z = null
z ||= 10
console.log z
z ||= 100
console.log z

console.log "&&= 判断两者都为 true 则赋值 ："
a= 1
b= 2
console.log a &&= b
console.log a
```

例 （源代码: assignment.js）

```
(function() {
var a, b, x, z;

console.log("= 赋值: ");

x = 10;

console.log(x);

console.log("+= 先加后赋值 : ");

x += 25;

console.log(x);

console.log("-= 先减后赋值 : ");

x -= 25;

console.log(x);

console.log("*= 先乘后赋值: ");

x *= 10;

console.log(x);

console.log("/= 先除后赋值: ");

x /= 10;

console.log(x);

console.log("%= 先取模后赋值: ");

x %= 3;

console.log(x);

console.log("?= 判断是否存在，不存在则赋值 : ");

if (typeof y === "undefined" || y === null) y = 3;

console.log(y);

if (typeof y === "undefined" || y === null) y = 100;

console.log(y);
```

```
console.log("||= 判断条件为假则赋值 : ");

z = null;

z || (z = 10);

console.log(z);

z || (z = 100);

console.log(z);

console.log("&&= 判断两者都为 true 则赋值: ");

a = 1;

b = 2;

console.log(a && (a = b));

console.log(a);

}).call(this);
```

输出 （源代码：assignment.coffee）

```
= 赋值 :
10
+= 先加后赋值 :
35
-= 先减后赋值 :
10
*= 先乘后赋值 :
100
/= 先除后赋值 :
10
%= 先取模后赋值:
1
?= 判断是否存在，不存在则赋值:
3
3
||= 判断条件为假则赋值 :
10
10
&&= 判断两者都为 true 则赋值:
2
2
```

同样，这些操作符都是直接映射的，很爽吧？

3.1.3　比较

下面，我们来看一下比较操作符，以及它们在 CoffeeScript 和 JavaScript 之间的映射情况。

- == 相等
- != 不等
- > 大于
- >= 大于等于
- < 小于
- <= 小于等于
- === 完全相同 （同类型且相等）
- !== 不完全相同

好，我们来看看这些操作符在 CoffeeScript 中是如何工作的。

例 （源代码：comparison.coffee）

```
console.log "== 相等：#{1 == 1}"

console.log "!= 不等：#{1 != 2}"

console.log "> 大于：#{2 > 1}"

console.log ">= 大于等于：#{1 >= 1}"

console.log "< 小于：#{1 < 2}"

console.log "<= 小于等于：#{1 < 2}"

console.log "=== 完全相同：#{'a' === 'a'}"

console.log "!== 不完全相同：#{1 !== 2}"
```

输出 （源代码：comparison.coffee）

```
Error: In content/control_structures/comparison.coffee, Parse error on line 13:
➥Unexpected '='
    at Object.parseError
➥ (/usr/local/lib/node_modules/coffee-script/lib/coffee-script/parser.js:470:11)
    at Object.parse
➥ (/usr/local/lib/node_modules/coffee-script/lib/coffee-script/parser.js:546:22)
    at /usr/local/lib/node_modules/coffee-script/lib/coffee-script/
➥ coffee-script.js:40:22
    at Object.run
➥ /usr/local/lib/node_modules/coffee-script/lib/coffee-script/coffee-script.js:68:34)
    at /usr/local/lib/node_modules/coffee-script/lib/coffee-script/command.js:135:29
```

```
  at /usr/local/lib/node_modules/coffee-script/lib/coffee-script/command.js:110:18
  at [object Object].<anonymous> (fs.js:114:5)
  at [object Object].emit (events.js:64:17)
  at afterRead (fs.js:1081:12)
  at Object.wrapper [as oncomplete] (fs.js:252:17)
```

是否觉得看着有点不对？我们来看看到底是怎么一回事。CoffeeScript 中不允许使用===和!==操作符。但愿，在印刷本书时，错误信息可以印刷得清楚明了。我仿佛能听到大家都在咆哮，咆哮为什么连这种最常用的比较操作符 CoffeeScript 都不支持，不过别担心，CoffeeScript 已经为我们想好了。且听我道来。

我们来重构下示例代码，这次把===和!==去掉。

例 （源代码：comparison2.coffee）

```
console.log "== 相等：#{1 == 1}"

console.log "!= 不等：#{1 != 2}"

console.log "> 大于：#{2 > 1}"

console.log ">= 大于等于：#{1 >= 1}"

console.log "< 小于：#{1 < 2}"

console.log "<= 小于等于：#{1 < 2}"
```

例 （源代码：comparison2.js）

```
(function() {

  console.log("== 相等：" + (1 === 1));

  console.log("!= 不等：" + (1 !== 2));

  console.log("> 大于：" + (2 > 1));

  console.log(">= 大于等于：" + (1 >= 1));

  console.log("< 小于：" + (1 < 2));

  console.log("<= 小于等于：" + (1 < 2));

}).call(this);
```

输出 （源代码：comparison2.coffee）

```
== 相等：true
!= 不等：true
> 大于：true
```

```
>= 大于等于：true
< 小于：true
<= 小于等于：true
```

太好了！我们的示例终于跑通了，不过，你应该已经注意到了一些有趣的东西。有没有发现==和!=被编译到 JavaScript 中后变成了什么？CoffeeScript 会分别将==和!=编译为===和!==。为什么会这样呢？让我们来看下在 JavaScript 中使用==和!=会发生什么。

例 （源代码：javascript_comparison.js）

```
x = 1;
y = '1';
console.log(x == y); // true
```

在上述例子中，尽管 1 和'1'完全是两个不同的对象，但它们还是相等。原因就是当使用==比较操作符时，JavaScript 会将比较对象转化为同一类型之后再进行比较。!=操作符也是如此。这是引发很多 JavaScript bug 的祸根。要想真正比较两个对象，就必须使用===操作符。

再来看下上述例子，这次用===操作符，你就会发现结果会是 false 而不是 true 了。

例 （源代码：javascript_comparison2.js）

```
x = 1;
y = '1';
console.log(x === y); // false
```

为了避免引发这类 bug，CoffeeScript 自动会将==和!=操作符转化为===和!==。是不是很赞？CoffeeScript 帮你将可能引发 bug 的风险降到了最低，你应当感谢它。另外，还有另一种使用===和!==操作符的方式，稍后讨论别名的时候会做相应介绍。

3.1.4 字符串

最后，还有一些和字符串相关的操作符：

- + 拼接
- += 拼接并赋值

下面是它们在 CoffeeScript 中的用法。

例 （源代码：string_operators.coffee）

```
console.log "+ 拼接: #{'a' + 'b'}"

x = 'Hello'
x += " World"
console.log "+= 拼接并赋值: #{x}"
```

例　（源代码：string_operators.js）

```
(function() {
  var x;

  console.log("+ 拼接: " + ('a' + 'b'));

  x = 'Hello';

  x += " World";

  console.log("+= 拼接并赋值: " + x);

}).call(this);
```

输出　（源代码：string_operators.coffee）

```
+ Concatenation: ab
+= Concatenate and assign: Hello World
```

幸运的是，这些操作符的用法和 JavaScript 中的一样。

3.1.5　与存在相关操作符

当我第一次发现 CoffeeScript 时，很快就被它的与存在相关操作符（existential operator）深深吸引了。该操作符让你使用简单的?就能检查一个变量或者函数是否存在。

我们来看一个简单的例子。

例　（源代码：existential1.coffee）

```
console.log x?
```

例　（源代码：existential1.js）

```
(function() {

  console.log(typeof x !== "undefined" && x !== null);

}).call(this);
```

输出　（源代码：existential1.coffee）

```
false
```

如上述代码所示，CoffeeScript 编译后的 JavaScript 会检查变量 x 是否有定义；如果定义了，则进一步检查该值是否为 null。这种方式书写条件语句非常强悍。

例（源代码：existential_if.coffee）

```
if html?
  console.log html
```

例（源代码：existential_if.js）

```
(function() {

  if (typeof html !== "undefined" && html !== null) console.log(html);

}).call(this);
```

上述例子仍旧对使用与存在相关操作符乐此不疲。用该操作符，我们可以检测对象（当然不止如此）是否存在，如果存在，可以调用该对象上的方法。关于这点，我最喜欢的例子就是检测 console 对象。如果你不了解 console 对象，可以把它理解为，在绝大多数浏览器端用于向 JavaScript 错误控制台输出消息的一种方式。通常，开发者会用它在特定的代码位置输出日志信息，以便于调试。我在本书中也大量使用 console 来展示示例代码的输出结果。

用 console 对象的问题就是，有些浏览器（Internet Explorer，说的就是你！）中并没有该对象。如果调用一个不存在的变量上的函数或者属性，浏览器会抛出异常，程序无法正常运行。与存在相关操作符就是为了帮助我们解决这类问题，比如，在某些浏览器（你知道我在说谁）上访问 console 对象。

我们先来看一个不用与存在相关操作符的例子。

例（源代码：existential2.coffee）

```
console.log "Hello, World"
console.log someObject.someFunction()
console.log "Goodbye, World"
```

例（源代码：existential2.js）

```
(function() {

  console.log("Hello, World");

  console.log(someObject.someFunction());

  console.log("Goodbye, World");

}).call(this);
```

输出（源代码：existential2.coffee）

```
Hello, World
ReferenceError: someObject is not defined
    at Object.<anonymous> (.../control_structures/existential2.coffee:5:15)
```

```
    at Object.<anonymous> (.../control_structures/existential2.coffee:9:4)
    at Module._compile (module.js:432:26)
    at Object.run
➥(/usr/local/lib/node_modules/coffee-script/lib/coffee-script/coffee-script.js:68:25)
    at /usr/local/lib/node_modules/coffee-script/lib/coffee-script/command.js:135:29
    at /usr/local/lib/node_modules/coffee-script/lib/coffee-script/command.js:110:18
    at [object Object].<anonymous> (fs.js:114:5)
    at [object Object].emit (events.js:64:17)
    at afterRead (fs.js:1081:12)
    at Object.wrapper [as oncomplete] (fs.js:252:17)
```

出错了，对吧？程序之所以挂了，是因为我们试图要调用的函数，它所属的对象根本不存在。下面，我们在使用前先用与存在相关操作符来做下检查，看看结果会怎样。

例（源代码：existential3.coffee）

```
console.log "Hello, World"
console.log someObject?.someFunction()
console.log "Goodbye, World"
```

例（源代码：existential3.js）

```
(function() {

  console.log("Hello, World");

  console.log(typeof someObject !== "undefined" && someObject !== null ?
➥someObject.someFunction() : void 0);

  console.log("Goodbye, World");

}).call(this);
```

输出（源代码：existential3.coffee）

```
Hello, World
undefined
Goodbye,World
```

确实好多了。尽管访问 someObject 变量时，输出了 undefined，但是程序正确执行完毕了。如果是实际场景下的例子的话，或许应该记录消息或者抛出警告，不过，目前这样使得我们的代码更加健壮了。

3.1.6　别名

为了让代码更加友好，CoffeeScript 对一些常用的操作符都定义了对应的别名。其中有些让 CoffeeScript 变得更加强大，有些则会引发歧义。表 3-1 列出了 CoffeeScript 中的别名以及对应的 JavaScript 中的操作符。

表 3-1　CoffeeScript 中的别名及其对应的 JavaScript 中的操作符

CoffeeScript	JavaScript
is	===
isnt	!==
not	!
and	&&
or	||
true, yes, on	true
false, no, off	false
@, this	this
of	in
in	N/A

我们来看看除最后两个之外的别名。最后两个别名会在第 5 章中介绍。

3.1.7 is 与 isnt 别名

下面是一个使用 is 和 isnt 操作符别名的例子。

例（源代码：is_aliases.coffee）

```
name = "mark"

console.log name is "mark"
console.log name isnt "bob"
```

例（源代码：is_aliases.js）

```
(function() {
  var name;

  name = "mark";

  console.log(name === "mark");

  console.log(name !== "bob");

}).call(this);
```

输出（源代码：is_aliases.coffee）

```
true
true
```

如上述代码所示，is 别名会被映射为===操作符，isnt 则会被映射为!==操作符。回想下本章前面所讨论过的，===和!==是 CoffeeScript 推荐使用的比较操作符。CoffeeScript 也同样鼓励使用 is 和 isnt 别名。它们有股“CoffeeScript 的味道”。此前有讨论过将==和!=这样的合法操作符从 CoffeeScript 中移除，强制使用 is 和 isnt 操作符。但一直到本书撰写时也还未实现，不过，至少你应该要有尽量多用 is 和 isnt，避免使用==和!=的意识。对于其他的操作符也是如此，都推荐使用别名来替代对应的操作符。

3.1.8 not 别名

我对 not 别名爱恨参半。我喜欢它的简洁，但讨厌它用起来总不如我意。not 别名和 JavaScript 中的!操作符一样；它会改变变量的布尔状态。它会让原本 true 的值变为 false，反之亦然。

我们来看下面 not 别名的用法的例子。

例（源代码：not_alias.coffee）

```
userExists = false

if not userExists
  console.log "the user doesn't exist!"
```

例（源代码：not_alias.js）

```
(function() {
  var userExists;

  userExists = false;

  if (!userExists) console.log("the user doesn't exist!");

}).call(this);
```

输出（源代码：not_alias.coffee）

```
the user doesn't exist!
```

如上述例子所示，在 not 别名和要改变布尔状态的变量之间要确保有一个空格。

那么，它让人讨厌的地方在哪里呢？在 CoffeeScript 中，有可能会写出如下代码。

例（源代码：not_alias_wrong.coffee）

```
name = "mark"

console.log name isnt "bob"
console.log name is not "bob"
```

例 （源代码：not_alias_wrong.js）

```
(function() {
  var name;

  name = "mark";

  console.log(name !== "bob");

  onsole.log(name === !"bob");

}).call(this);
```

输出 （源代码：not_alias_wrong.coffee）

```
true
false
```

尽管从英语语法上看上述那两行代码是一样的，但事实上，这两行代码是完全不同的。使用 isnt 别名的那行代码检查的是两个对象是否不等，而用 not 别名的那行代码检查的则是第一个变量与第二个变量相反的布尔值是否相等。这个错误很容易犯，特别是刚刚使用 CoffeeScript 的时候。

3.1.9 and 与 or 别名

我喜欢 and 和 or 这两个别名，它们不仅让代码更具可读性，而且用起来没有歧义。来看下面这个例子。

例 （源代码：and_or.coffee）

```
if true and true
  console.log "true and true really is true"

if false or true
  console.log "something was true"
```

例 （源代码：and_or.js）

```
(function() {

  if (true && true) console.log("true and true really is true");

  if (false || true) console.log("something was true");

}).call(this);
```

输出（源代码：and_or.coffee）

```
true and true really is true
something was true
```

如我所述，它们都工作得很好。

3.1.10　布尔相关的别名

CoffeeScript 不仅支持 true 和 false 这类布尔相关的别名，还借鉴了 YAML[①]并且添加了一些其他的别名来让代码更具可读性。

我们来看看下面的例子。

例（源代码：boolean_operators.coffee）

```
myAnswer = true
console.log myAnswer is yes
console.log myAnswer is true

light = true
console.log light is on
console.log light is true

myAnswer = false
console.log myAnswer is no
console.log myAnswer is false

light = false
console.log light is off
console.log light is false
```

例（源代码：boolean_operators.js）

```
(function() {
  var light, myAnswer;

  myAnswer = true;

  console.log(myAnswer === true);

  console.log(myAnswer === true);

  light = true;

  console.log(light === true);
```

① http://www.yaml.org/spec/1.2/spec.html

```
  console.log(light === true);

  myAnswer = false;

  console.log(myAnswer === false);

  console.log(myAnswer === false);

  light = false;

  console.log(light === false);

  console.log(light === false);

}).call(this);
```

输出 （源代码：boolean_operators.coffee）

```
true
true
true
true
true
true
true
true
```

如上述例子所示，yes、no、on 以及 off 都是 CoffeeScript 中的别名，可以让代码变得更有趣也更具可读性。

3.1.11　@别名

要给大家介绍的最后一个别名是@。本书中会在介绍 CoffeeScript 不同内容时多次提到该别名。但这里我们介绍的是其最基础也是最常见的用法：用作 JavaScript 中 this 关键字的别名。

下面是一个使用@别名的非常简单的例子。

例 （源代码：at_alias.coffee）

```
object = {
  name: 'mark'
  sayHi: ->
    console.log "Hello: #{@name}"
}
```

```
object.sayHi()

console.log @name
```

例（源代码：`at_alias.js`）

```
(function() {
  var object;

  object = {
    name: 'mark',
    sayHi: function() {
      return console.log("Hello: " + this.name);
    }
  };

  object.sayHi();

  console.log(this.name);

}).call(this);
```

输出（源代码：`at_alias.coffee`）

```
Hello: mark
undefined
```

如果不清楚上述 JavaScript 例子中 `this` 的作用域甚至根本不知道 `this` 是什么，那么请务必把本书暂放一旁，找一本合适的 JavaScript 书先读一下。

上述例子中，如果对比 CoffeeScript 和编译后的 JavaScript 代码，就会发现所有 CoffeeScript 中的`@`在编译后的代码中都被替换成了 `this`。我发现代码中用了`@`之后，可读性更强了。可以很容易地区分函数中的“局部”变量（有时也叫做实例变量）和当前函数外定义的变量和函数。

3.2 if/unless

在我作为开发者的历程中，还未见过一门编程语言中没有条件语句这一概念的。当然了，肯定有人在尝试着写这样一门语言，不过我感觉这样的语言有点不靠谱。

条件语句使得程序更加智能，它能让应用对不同的场景做出响应。用户是否登录？如果还没有，就要求他登录；否则，展示与其个人相关的页面。当前用户是否已付款？是，则允许他访问账号；否则，告诉他去付款。这些都是条件语句的例子。程序会根据这些问题的不同答案执行不同路径的代码。

和大多数其他编程语言一样，CoffeeScript 也提供了条件语句。这些条件语句通常和我们在本章中介绍过的操作符以及别名配合使用，来帮助程序作出更加智能的决策。

3.2.1 `if` 语句

本书中提到过 if 语句的例子，其结构很简单。

例 （源代码：if.coffee）

```
if true
  console.log "the statement was true"
```

例 （源代码：if.js）

```
(function() {

  if (true) console.log("the statement was true");

}).call(this);
```

输出 （源代码：if.coffee）

```
the statement was true
```

上述例子尽管有点难以置信地做作，不过展示了 if 语句的结构。不论条件语句是什么，都先使用 if 关键字。如果条件语句返回 true，则执行 if 的代码块；如果条件语句返回 false，则跳过该代码块。

来看一个稍微不那么做作的例子。

例 （源代码：if2.coffee）

```
today = "Sunday"
if today is "Sunday"
  console.log "Today is Sunday"
```

例 （源代码：if2.js）

```
(function() {
  var today;

  today = "Sunday";

  if (today === "Sunday") console.log("Today is Sunday");

}).call(this);
```

输出 （源代码：if2.coffee）

```
Today is Sunday
```

3.2.2 if/else 语句

有时我们想要在条件为真的时候执行一些代码，有时我们也想要在条件为假的时候，执行另外一些代码。这些情况下，我们就可以使用 if 语句以及 else 关键字来定义当条件为假时要执行的代码块。

下面来看一个例子。

例（源代码：if_else.coffee）

```
today = "Monday"
if today is "Sunday"
  console.log "Today is Sunday"
else
  console.log "Today is not Sunday"
```

例（源代码：if_else.js）

```
(function() {
  var today;

  today = "Monday";

  if (today === "Sunday") {
    console.log("Today is Sunday");
  } else {
    console.log("Today is not Sunday");
  }

}).call(this);
```

输出（源代码：if_else.coffee）

```
Today is not Sunday
```

上述例子所示，因为条件为假，所以定义在 else 关键字后的代码块就会被执行。

CoffeeScript 支持将整个 if/else 语句写在一行里，如下所示：

例（源代码：if_else_1_line.coffee）

```
today = "Monday"
console.log if today is "Sunday" then "Today is Sunday" else "Today is not Sunday"
```

例（源代码：if_else_1_line.js）

```
(function() {
  var today;

  today = "Monday";
```

```
  console.log(today === "Sunday" ? "Today is Sunday" : "Today is not Sunday");

}).call(this);
```

输出 (源代码: if_else_1_line.coffee)

```
Today is not Sunday
```

> **提示**：个人觉得，尽管将 if、else、then 语句写在一行字数变少了，但是可读性却变差了，所以我很少这样写。

在 JavaScript 中可以使用所谓的三元操作符[1]将上述例子写成一行代码，CoffeeScript 编译后的代码就是如此。但 CoffeeScript 并不支持三元操作符，使用它会编译出很奇怪的 JavaScript 代码。下面我们来看一下。

例 (源代码: ternary.coffee)

```
today = "Monday"
console.log today is "Sunday" ? "Today is Sunday" : "Today is not Sunday"
```

例 (源代码: ternary.js)

```
(function() {
  var today, _ref;

  today = "Monday";

  console.log((_ref = today === "Sunday") != null ? _ref : {
    "Today is Sunday":  "Today is not Sunday"
  });

}).call(this);
```

输出 (源代码: ternary.coffee)

```
false
```

我不打算解释上述代码了。建议你在此做一下标记，读完本书后再回来看看，是否能够找出为什么 CoffeeScript 会编译成这样的 JavaScript。

3.2.3 if/else if 语句

暂时假设我们要写一个很简单的待办事宜应用。其中有这样一个功能：如果今天是周六，

① http://en.wikipedia.org/wiki/Ternary_operation

应用应该展示当天的待办事宜；如果是周日，应用应该提醒我休息，好好享受一天；最后，如果既不是周六也不是周日，应用应该提醒我要去工作了。如何使用我们已掌握的知识实现该功能呢？请看如下代码。

例（源代码：if_else_if_1.coffee）

```
today = "Monday"
if today is "Saturday"
  console.log "Here are your todos for the day..."
if today is "Sunday"
  console.log "Go watch football and relax!"
if today isnt "Saturday" and today isnt "Sunday"
  console.log "Get to work you lazy bum!"
```

例（源代码：if_else_if_1.js）

```
(function() {
  var today;

  today = "Monday";

  if (today === "Saturday") console.log("Here are your todos for the day...");

  if (today === "Sunday") console.log("Go watch football and relax!");

  if (today !== "Saturday" && today !== "Sunday") {
    console.log("Get to work you lazy bum!");
  }

}).call(this);
```

输出（源代码：if_else_if_1.coffee）

```
Get to work you lazy bum!
```

尽管上述代码可以工作，但却不是最高效的。首先，它会检查每一个 `if` 语句，哪怕第一个条件为 `true`，也还是会检查其他的 `if` 语句，看看是不是也为 `true`，尽管事实上这根本是不可能的。其次，最后一个 `if` 语句多少有点重复，因为前面已经问过类似的问题了。最后，它也很容易出错。如果我们使用缩写而不是全名，如用 Sun 来表示 Sunday，那不得不将所有的 Sunday 都替换成 Sun。

使用 `else if` 语句就可以很好地解决这些问题，下面给出具体的做法。

例（源代码：if_else_if_2.coffee）

```
today = "Monday"
if today is "Saturday"
  console.log "Here are your todos for the day..."
```

```
else if today is "Sunday"
  console.log "Go watch football and relax!"
else
  console.log "Get to work you lazy bum!"
```

例（源代码：if_else_if_2.js）

```
(function() {
  var today;

  today = "Monday";

  if (today === "Saturday") {
    console.log("Here are your todos for the day...");
  } else if (today === "Sunday") {
    console.log("Go watch football and relax!");
  } else {
    console.log("Get to work you lazy bum!");
  }

}).call(this);
```

输出（源代码：if_else_if_2.coffee）

```
Get to work you lazy bum!
```

现在是不是看上去好很多了？而且代码也更高效了。如果今天是周六，会执行第一个代码块，并跳过其余的 else if 和 else 语句，因为没有必要再去检查了。

本章后续部分会介绍使用 switch 语句来改写上述例子，它会让代码更加整洁。

3.2.4 unless 语句

Ruby 语言中有 unless 语句的概念，CoffeeScript 的创造者们觉得这是个很棒的概念，也借鉴了过来。

unless 语句是用来干嘛的呢？简单说来，它可以让 else 放在 if 的前面。我知道你现在肯定在挠头，尽管一开始会比较困惑，不过它其实很直观。

使用 unless 时，我们检查条件语句的值是否为假。如果为假，则执行对应的代码块。下面举个例子。

例（源代码：unless.coffee）

```
today = "Monday"
unless today is "Sunday"
  console.log "No football today!"
```

例（源代码：unless.js）

```
(function() {
  var today;

  today = "Monday";

  if (today !== "Sunday") console.log("No football today!");

}).call(this);
```

输出（源代码：unless.coffee）

```
No football today!
```

上述例子也可以改写为如下形式。

例（源代码：unless_as_if.coffee）

```
today = "Monday"
unless today is "Sunday"
  console.log "No football today!"

if not (today is "Sunday")
  console.log "No football today!"

if today isnt "Sunday"
  console.log "No football today!"
```

例（源代码：unless_as_if.js）

```
(function() {
  var today;

  today = "Monday";

  if (today !== "Sunday") console.log("No football today!");

  if (!(today === "Sunday")) console.log("No football today!");

  if (today !== "Sunday") console.log("No football today!");

}).call(this);
```

输出（源代码：unless_as_if.coffee）

```
No football today!
No football today!
No football today!
```

在上述三个例子中，我个人最喜欢最后一个使用 isnt 别名的例子，我觉得这种方式代码更整洁、可读性更强。不过，选择哪一种最终取决于你，这三种方式都是正确的。

3.2.5 内联条件语句

除了 unless 关键字外，CoffeeScript 团队还借鉴了 Ruby 中的内联条件语句的想法。内联条件语句允许将条件语句以及对应的代码块写在一行里。

下面的例子展示了其用法。

例 （源代码：inline.coffee）

```
today = "Sunday"
console.log "Today is Sunday" if today is "Sunday"
```

例 （源代码：inline.js）

```
(function() {
  var today;

  today = "Sunday";

  if (today === "Sunday") console.log("Today is Sunday");

}).call(this);
```

输出 （源代码：inline.coffee）

```
Today is Sunday
```

和本章前面介绍的与存在相关操作符一起使用时，内联条件语句可以保持代码的整洁。

3.3 switch/case 语句

前面在介绍 else if 语句时，我提过可以使用 switch 语句[①]来让代码更整洁。switch 语句可以构建一张条件表，用来后续匹配对象。当匹配其中一个条件时，对应的代码就会被执行。当一个都匹配不到，我们也可以提供 else 分支来指定这种情况下要执行的代码。

我们将此前介绍 else if 语句时的例子用 switch 语句做了改写。

例 （源代码：switch1.coffee）

```
today = "Monday"
switch today
  when "Saturday"
    console.log "Here are your todos for the day..."
```

① http://en.wikipedia.org/wiki/Switch_statement

```
  when "Sunday"
    console.log "Go watch football and relax!"
  else
    console.log "Get to work you lazy bum!"
```

例（源代码：switch1.js）

```
(function() {
  var today;

  today = "Monday";

  switch (today) {
    case "Saturday":
      console.log("Here are your todos for the day...");
      break;
    case "Sunday":
      console.log("Go watch football and relax!");
      break;
    default:
      console.log("Get to work you lazy bum!");
  }

}).call(this);
```

输出（源代码：switch1.coffee）

```
Get to work you lazy bum!
```

如上述例子所示，today 变量会依次与定义好的 case 语句进行匹配。如果 today 变量与任何 case 语句都不匹配，程序会跳转到 switch 语句结尾定义的 else 语句的位置来继续执行。值得指出的是，结尾可以不定义 else 代码块，这样的话，上述例子就不会有任何输出。

善于观察的你肯定已经发现了，在编译后的 JavaScript 代码中，每个 case 语句最后都会有一个 break 关键字。JavaScript 中允许 switch 语句按顺序去匹配多个 case 语句。不过，大多数情况下不推荐这样做，因为这样经常会引发很多 bug。最常见的 bug 就是当匹配了一种情况之后，执行了对应的代码，最后还会去执行 default 代码块中的代码。这里，break 关键字就是让 switch 语句停止执行，不再去匹配其他的情况。这又是 CoffeeScript 用心良苦的一个例子，它能帮助避免此类 bug 的发生。

switch 语句还支持用逗号分隔的值列表来匹配 case 关键字。它可以为多个匹配执行同一代码块。比方说，我想要检查今天是否是周末，如果是，则提醒我休息，否则，提醒我工作。这样的情况就可以采用如下这种方式来实现。

例（源代码：switch2.coffee）

```
today = "Sunday"
switch today
  when "Saturday", "Sunday"
```

```
    console.log "Enjoy your #{today}!"
  else
    console.log "Off to work you go. :("
```

例（源代码：switch2.js）

```
(function() {
  var today;

  today = "Sunday";

  switch (today) {
    case "Saturday":
    case "Sunday":
      console.log("Enjoy your " + today + "!");
      break;
    default:
      console.log("Off to work you go. :(");
  }

}).call(this);
```

输出（源代码：switch2.coffee）

```
Enjoy your Sunday!
```

至此，对 switch 语句的介绍就差不多了。对于什么时候用 switch 语句，在哪里用，甚至该不该用 switch 语句，开发者中有很多争论。尽管我不会天天用它，不过它绝对有它存在的意义。因此，什么时候用由你根据自己的应用而定。

3.4　小结

本章介绍了很多基础内容，我们介绍了 CoffeeScript 中的各种操作符、其使用方法以及对应的 JavaScript 的操作符，介绍了 CoffeeScript 帮助我们书写更好 JavaScript 代码的地方，还介绍了让代码读起来更像人类语言而非计算机语言的别名。另外还有如何构建条件语句来让程序更加智能，能够根据特定的条件正确执行对应的代码。最后，我们介绍了如何使用 switch 语句来简化复杂的代码。

在本书最初的大纲中，本章属于第 2 章的一部分，但我个人感觉这部分涵盖了很多内容，可以独立成为一章。本章本应叫做“基础知识 —— 第 2 部分”。之所以我要说这些，是要告诉你，掌握了本章和第 2 章的知识，就掌握了 CoffeeScript 的基本构件块。现在我们可以介绍一些真正有意思的东西了。

第 4 章

函数与参数

本章开始介绍编程语言中最重要的部分——函数。函数封装可重用的离散代码块。如果没有它，代码将会变得冗长、可读性差且难以维护。

我很想给你举一个在 JavaScript 中不能用函数也不能写函数的例子，但是我真的做不到。因为在 JavsScript 中，哪怕是最简单的将一个字符串变成小写的例子都需要用到函数。

我无法给出一个不用函数的例子，但是我能给出一个使用一两个函数的 CoffeeScript 的例子，以示函数对于提高代码的可控性是多么重要。

例（源代码：no_functions_example.coffee）

```
tax_rate = 0.0625

val = 100
console.log "What is the total of $#{val} worth of shopping?"
tax = val * tax_rate
total = val + tax
console.log "The total is #{total}"

val = 200
console.log "What is the total of $#{val} worth of shopping?"
tax = val * tax_rate
total = val + tax
console.log "The total is #{total}"
```

例（源代码：no_functions_example.js）

```
(function() {
  var tax, tax_rate, total, val;

  tax_rate = 0.0625;
```

```
  val = 100;

  console.log("What is the total of $" + val + " worth of shopping?");

  tax = val * tax_rate;

  total = val + tax;

  console.log("The total is " + total);

  val = 200;

  console.log("What is the total of $" + val + " worth of shopping?");

  tax = val * tax_rate;

  total = val + tax;

  console.log("The total is " + total);

}).call(this);
```

输出 （源代码：no_functions_example.coffee）

```
What is the total of $100 worth of shopping?
The total is 106.25
What is the total of $200 worth of shopping?
The total is 212.5
```

上述例子计算的是购买含税物品的总价。例子虽然简单，但能看到其中为了多次计算含税总价，重复使用了一段代码。

下面我们采用函数来重构上述代码，试着让代码更整洁。

例 （源代码：with_functions_example.coffee）

```
default_tax_rate = 0.0625

calculateTotal = (sub_total, rate = default_tax_rate) ->
  tax = sub_total * rate
  sub_total + tax

val = 100
console.log "What is the total of $#{val} worth of shopping?"
console.log "The total is #{calculateTotal(val)}"

val = 200
console.log "What is the total of $#{val} worth of shopping?"
console.log "The total is #{calculateTotal(val)}"
```

例（源代码：with_functions_example.js）

```
(function() {
  var calculateTotal, default_tax_rate, val;

  default_tax_rate = 0.0625;

  calculateTotal = function(sub_total, rate) {
    var tax;
    if (rate == null) rate = default_tax_rate;
    tax = sub_total * rate;
    return sub_total + tax;
  };

  val = 100;

  console.log("What is the total of $" + val + " worth of shopping?");

  console.log("The total is " + (calculateTotal(val)));

  val = 200;

  console.log("What is the total of $" + val + " worth of shopping?");

  console.log("The total is " + (calculateTotal(val)));

}).call(this);
```

输出（源代码：with_functions_example.coffee）

```
What is the total of $100 worth of shopping?
The total is 106.25
What is the total of $200 worth of shopping?
The total is 212.5
```

上述例子中有些你可能看不懂，不过不用担心，这都是本章要介绍的。不过即使你不知道在 CoffeeScript 中函数是如何定义的以及工作的，你应该也能看到重构后的例子中代码明显整洁多了。在重构的代码中，甚至还支持任意不同的税率，这也是我们需要的。函数有助于让代码保持 DRY[①]原则：不重复自己（Don't Repeat Yourself）。DRY 原则让代码更易管理、bug 更少。

4.1　函数基础

我们开始介绍在 CoffeeScript 中如何定义函数的基础知识。下面展示了一个非常简单的函数的基本结构。

① http://en.wikipedia.org/wiki/DRY

例（源代码：simple_function.coffee）

```
myFunction = ()->
  console.log "do some work here"

myFunction()
```

例（源代码：simple_function.js）

```
(function() {
  var myFunction;

  myFunction = function() {
    return console.log("do some work here");
  };

  myFunction();

}).call(this);
```

上述例子中，函数有一个名字（myFunction）[1]，其后有一个相应的代码块。其函数体是在->后缩进的代码块，缩进规则遵循我们第 2 章——“基础知识”中介绍的有意义的空格规则。

该函数不接受任何参数，因为->前是一对空括号。在 CoffeeScript 中调用一个无参函数时，要求使用括号，如 myFunction()。

在定义无参函数时，完全可以把括号去掉，如下所示。

例（源代码：simple_function_no_parens.coffee）

```
myFunction = ->
  console.log "do some work here"

myFunction()
```

例（源代码：simple_function_no_parens.js）

```
(function() {
  var myFunction;

  myFunction = function() {
    return console.log("do some work here");
  };

  myFunction();

}).call(this);
```

① 这里名字的意思是，将函数赋值给 myFunction 变量。——译者注

还有另外一种书写简单函数的方式。因为上述例子中的函数体代码只有一行，所以我们可以将整个函数定义都写在一行上，如下所示。

例（源代码：simple_function_one_line.coffee）

```
myFunction = -> console.log "do some work here"

myFunction()
```

例（源代码：simple_function_one_line.js）

```
(function() {
  var myFunction;

  myFunction = function() {
    return console.log("do some work here");
  };

  myFunction();

}).call(this);
```

上述三个例子代码都是等效的，并且调用方式也是一样的。

> **提示**：尽管可以用一行代码完成此类函数定义，但这种方式我并不推荐。个人觉得，这种方式并不会让代码更整洁或是更易读。并且，将函数体写在单独的代码行中，可以使后期增加函数体代码变得更方便。
> 你还应该注意到每个函数的最后一行都包含一个 return 关键字，这是 CoffeeScript 自动加上的。不论函数最后一行是什么，都会成为函数的返回值。这和 Ruby 这类语言的处理方式是一样的。因为 CoffeeScript 会自动在编译后的 JavaScript 代码中加入 return 关键字，所以 return 的使用在 CoffeeScript 中是可选的。

> **提示**：我发现有时显式地写上 return 会让代码的意思更清楚，所以建议在觉得会有助于代码可读性的地方加上 return。

> **提示**：如果不想将函数的最后一行成为函数的返回值，需要显式地另起一行，写上 return 关键字。比方说，return null 或者 return undefined 之类的。

4.2　参数

和在 JavaScript 中一样，CoffeeScript 中的函数也可以接收参数。参数可以将对象传递给函数，然后函数内部可以对其进行计算、数据操作等任何想要做的处理。

在 CoffeeScript 中，定义有参数的函数和在 JavaScript 中没什么两样。在函数定义的括号中，用逗号来分隔函数接收的参数名。

例 （源代码：function_with_args.coffee）

```
calculateTotal = (sub_total, rate) ->
  tax = sub_total * rate
  sub_total + tax

console.log calculateTotal(100, 0.0625)
```

例 （源代码：function_with_args.js）

```
(function() {
  var calculateTotal;

  calculateTotal = function(sub_total, rate) {
    var tax;
    tax = sub_total * rate;
    return sub_total + tax;
  };

  console.log(calculateTotal(100, 0.0625));

}).call(this);
```

输出 （source：function_with_args.coffee）

```
106.25
```

上述例子中，定义了一个函数，这个函数接收两个参数，并对其做数学运算计算出含税物品的总价。在调用该函数时，我们传递了两个值进去。

第 2 章中，我们简单介绍了 CoffeeScript 中括号的规则。这里我再重申下其中一条规则。对于接收参数的函数，在调用的时候可以忽略括号。也就是说，我们可以将上述例子写成如下形式。

例 （源代码：function_with_args_no_parens.coffee）

```
calculateTotal = (sub_total, rate) ->
  tax = sub_total * rate
  sub_total + tax

console.log calculateTotal 100, 0.0625
```

例 （源代码：function_with_args_no_parens.js）

```
(function() {
  var calculateTotal;

  calculateTotal = function(sub_total, rate) {
    var tax;
    tax = sub_total * rate;
```

```
    return sub_total + tax;
  };

  console.log(calculateTotal(100, 0.0625));

}).call(this);
```

输出（源代码：function_with_args_no_parens.coffee）

```
106.25
```

如上述例子所示，CoffeeScript会在编译后的JavaScript代码中自动加上这些括号。

> **提示**：使用括号调用函数在CoffeeScript界引发了热议。我个人倾向于使用括号。我觉得这样不但会让代码可读性更好，而且能避免因为编译器误放括号而引起的潜在bug。如果你对是否使用括号举棋不定，那就使用括号调用函数。这样你一定不会后悔的。

4.3 默认参数值

在像Ruby这样的编程语言中，可以给参数赋予默认值。这意味着，不论什么原因，只要没有传递参数值，就可以使用合理的默认参数值。

我们再来看一下计算含税物品总价的例子。这次我们给税率参数rate赋予一个默认的值，以供没有传入该参数时使用。

例（源代码：default_args.coffee）

```
calculateTotal = (sub_total, rate = 0.05) ->
  tax = sub_total * rate
  sub_total + tax

console.log calculateTotal 100, 0.0625
console.log calculateTotal 100
```

例（源代码：default_args.js）

```
(function() {
  var calculateTotal;

  calculateTotal = function(sub_total, rate) {
    var tax;
    if (rate == null) rate = 0.05;
    tax = sub_total * rate;
    return sub_total + tax;
  };

  console.log(calculateTotal(100, 0.0625));
```

```
  console.log(calculateTotal(100));

}).call(this);
```

输出（源代码：default_args.coffee）

```
106.25
105
```

在上述例子中，我们在 CoffeeScript 代码中给 rate 参数设置了默认值 0.05。第一次调用 calculateTotal 函数时，我们传递的 rate 参数值为 0.0625；第二次调用时，我们省略了 rate 参数，函数内部则使用了其默认值 0.05。

我们再来进一步看看默认参数值，这次将其指向其他参数值。来看下面这个例子。

例（源代码：default_args_referring.coffee）

```
href = (text, url = text) ->
  html = "<a href='#{url}'>#{text}</a>"
  return html

console.log href("Click Here", "http://www.example.com")
console.log href("http://www.example.com")
```

例（源代码：default_args_referring.js）

```
(function() {
  var href;

  href = function(text, url) {
    var html;
    if (url == null) url = text;
    html = "<a href='" + url + "'>" + text + "</a>";
    return html;
  };

  console.log(href("Click Here", "http://www.example.com"));

  console.log(href("http://www.example.com"));

}).call(this);
```

输出（源代码：default_args_referring.coffee）

```
<a href='http://www.example.com'>Click Here</a>
<a href='http://www.example.com'>http://www.example.com</a>
```

上述例子中，在不传递 url 参数的情况下，其值会被设置为传入的 text 参数值。

默认参数值还可以是函数。由于只有在未传递参数时，默认的参数值才会被调用，因此函数作为默认参数值也不会有性能问题。

例（源代码：default_args_with_function.coffee）

```
defaultRate = -> 0.05

calculateTotal = (sub_total, rate = defaultRate()) ->
  tax = sub_total * rate
  sub_total + tax

console.log calculateTotal 100, 0.0625
console.log calculateTotal 100
```

例（源代码：default_args_with_function.js）

```
(function() {
  var calculateTotal, defaultRate;

  defaultRate = function() {
    return 0.05;
  };

  calculateTotal = function(sub_total, rate) {
    var tax;
    if (rate == null) rate = defaultRate();
    tax = sub_total * rate;
    return sub_total + tax;
  };

  console.log(calculateTotal(100, 0.0625));

  console.log(calculateTotal(100));

}).call(this);
```

输出（源代码：default_args_with_function.coffee）

```
106.25
105
```

> **提示**：使用默认参数值时，很重要的一点就是它们必须要放在参数列表的最后。可以为多个参数设置默认参数值，但都必须要放在参数列表的最后。

4.4 splat 操作符

有时定义函数时，我们不确定该函数要接收多少个参数。有时需要一个，有时可能需要很多。怎么办呢？为了解决这个问题，CoffeeScript 允许在为函数定义参数列表时使用可选的 splat

操作符。splat 是指在方法定义后的括号中使用...。

> **提示**：有一种很好的方法能帮助记忆如何使用 splat 操作符，这种方法就是采用 etc...。这种方式既好记，又很酷。

什么时候使用 splat 操作符呢？在函数接收可变数目的参数时就可以使用。在看详细案例前，先来看一个使用 splat 操作符的简单例子。

例（源代码：splats.coffee）

```
splatter = (etc...) ->
  console.log "Length: #{etc.length}, Values: #{etc.join(', ')}"

splatter()
splatter("a", "b", "c")
```

例（源代码：splats.js）

```
(function() {
  var splatter,
    __slice = Array.prototype.slice;

  splatter = function() {
    var etc;
    etc = 1 <= arguments.length ? __slice.call(arguments, 0) : [];
    return console.log("Length: " + etc.length + ", Values: " + (etc.join(', ')));
  };

  splatter();

  splatter("a", "b", "c");

}).call(this);
```

输出（源代码：splats.coffee）

```
Length: 0, Values:
Length: 3, Values: a, b, c
```

如上述例子所示，传入的参数会被自动转化为一个数组，没有参数传入时，则转化为空数组。

> **提示**：splat 操作符很强大，但在 JavaScript 中实现需要很多额外的代码。看看上面的例子，编译后的 JavaScript 中相对应的实现 splat 操作符的部分，想必你肯定不愿意去写这些额外的代码。

不像其他语言对 splat 只支持一种固定的格式，CoffeeScript 不强制要求 splat 操作符一定要放在参数列表的最后。事实上，它可以出现在参数列表的任意位置。不过，要注意的是，参数列表中只能有一个 splat 操作符。

为了说明 splat 如何用在参数列表中的任意部分，我们写一个将接收的参数转化为字符串的方法。在构建该字符串时，我们确保第一个参数和最后一个参数都是大写的，其他参数是小写的。然后，用斜杠将这些参数以字符串形式拼接起来。

例（源代码: splats_arg_join.coffee）

```
joinArgs = (first, middles..., last) ->
  parts = []

  if first?
    parts.push first.toUpperCase()

  for middle in middles
    parts.push middle.toLowerCase()

  if last?
    parts.push last.toUpperCase()

  parts.join('/')

console.log joinArgs("a")
console.log joinArgs("a", "b")
console.log joinArgs("a", "B", "C", "d")
```

例（源代码: splats_arg_join.js）

```
(function() {
  var joinArgs,
    __slice = Array.prototype.slice;

  joinArgs = function() {
    var first, last, middle, middles, parts, _i, _j, _len;
    first = arguments[0], middles = 3 <= arguments.length ? __slice.call(arguments, 1,
➥_i = arguments.length - 1) : (_i = 1, []), last = arguments[_i++];
    parts = [];
    if (first != null) parts.push(first.toUpperCase());
    for (_j = 0, _len = middles.length; _j < _len; _j++) {
      middle = middles[_j];
      parts.push(middle.toLowerCase());
    }
    if (last != null) parts.push(last.toUpperCase());
    return parts.join('/');
  };

  console.log(joinArgs("a"));

  console.log(joinArgs("a", "b"));
```

```
  console.log(joinArgs("a", "B", "C", "d"));

}).call(this);
```

输出　（源代码：splats_arg_join.coffee）

```
A
A/B
A/b/c/D
```

我承认上述例子稍显复杂，不过它很好地说明了 splat 操作符的用法。当调用 joinArgs 函数时，传入函数的第一个参数会被赋值给 first 变量，最后一个参数会被赋值给 last 变量，而第一个参数与最后一个参数之间如果还有变量的话会被统一放入一个数组中，并将该数组赋值给 middles 变量。

> **提示：**我们也可以将函数设计为只接收一个 splat 参数，随后再从 middles 数组中解析出第一个元素和最后一个元素，不过，上述例子中这种方式好在我们可以不用写解析代码。真爽。

最后，使用 splat 时，有时可能想要传递一个数组作为一个单独的参数，这也是可以的。来看一个简单例子。

例　（源代码：splats_array.coffee）

```
splatter = (etc...) ->
  console.log "Length: #{etc.length}, Values: #{etc.join(', ')}"

a = ["a", "b", "c"]
splatter(a)
splatter(a...)
```

例　（源代码：splats_array.js）

```
(function() {
  var a, splatter,
    __slice = Array.prototype.slice;

  splatter = function() {
    var etc;
    etc = 1 <= arguments.length ? __slice.call(arguments, 0) : [];
    return console.log("Length: " + etc.length + ", Values: " + (etc.join(', ')));
  };

  a = ["a", "b", "c"];

  splatter(a);
```

```
  splatter.apply(null, a);

}).call(this);
```

输出 (源代码: `splats_array.coffee`)

```
Length: 1, Values: a,b,c
Length: 3, Values: a, b, c
```

上述例子中，我们将数组作为参数传入，不过，正如你看到的，splatter 函数会将该数组作为一个参数，事实也是如此，它本来就是一个参数。然而，如果在传入的数组后加上...，CoffeeScript 会将该数组内的元素分割出来作为一个个单独参数传递给调用函数。

4.5 小结

本章介绍了所有你想要知道的关于 CoffeeScript 中函数的所有内容。我们一开始介绍了如何定义简单函数的方法，事实上是介绍了多种在 CoffeeScript 中定义函数的方式，随后介绍了如何定义函数的参数以及如何调用函数，包括何时该在调用函数时使用括号，还介绍了 CoffeeScript 中我最喜欢的特性之一 —— 默认参数值。

最后，介绍了 splat 操作符，以及它是如何帮助我们写接收可变数目的参数的函数的。

掌握了本章函数与参数的内容后，我们可以进入第 5 章了。准备好了吗？

第 5 章

集合与迭代

集合（collection）在几乎所有的面向对象编程语言中都是很重要的一部分。集合能让我们很容易地将多个值存储在如数组这样的列表中，或者使用像 JavaScript 中对象这样的键/值对的形式来进行存储。我们可以用对象来表示某个事物，如一本书，然后将标题、作者以及出版日期这样的值赋给该对象。数组则可以用来存储创建的书对象列表。

讲到集合就一定要提迭代。迭代允许对存储的数据，如书籍列表，进行遍历，做诸如：逐个打印到屏幕上、对每本书的信息进行相应的更新，甚至是任何想要做的事。

本章前半部分介绍 CoffeeScript 中的数组和对象。我们还会看看它们对应到 JavaScript 中是怎样的。因为是 CoffeeScript，自然还少不了介绍它里面关于数组和对象的有趣特性。

本章后半部分集中介绍迭代。结合前半部分的集合，我们介绍如何遍历和操作集合。

5.1 数组

这里不讲诸如数组在内存中是如何实现的这样的话题，作为阅读本书的前提，我们都应当知道数组是简单的数据结构，用来以连续列表的形式存储数据。除了特别处理，默认情况下，新的数据项都会存储到列表的末尾。CoffeeScript 中的数组和 JavaScript 中的数组并没有什么不同，数组下标都是从零开始的，构造数组的方式也相同。

例（源代码：array1.coffee）

```
myArray = ["a", "b", "c"]

console.log myArray
```

例（源代码：array1.js）

```
(function() {
  var myArray;
```

```
  myArray = ["a", "b", "c"];

  console.log(myArray);

}).call(this);
```

输出　（源代码：array1.coffee）

```
[ 'a', 'b', 'c' ]
```

在 CoffeeScript 和 JavaScript 中，除了前者可以不写分号和 var 关键字之外，它们在数组的实现上几乎相同。不过，CoffeeScript 还是针对数组提供了一些有意思的特性。

在 CoffeeScript 中，定义数组时，可以不用逗号来分隔数组项，只需简单地将每个数组项都单独写在一行即可。

例　（源代码：array2.coffee）

```
myArray = [
            "a"
            "b"
            "c"
          ]

console.log myArray
```

例　（源代码：array2.js）

```
(function() {
  var myArray;

  myArray = ["a", "b", "c"];

  console.log(myArray);

}).call(this);
```

输出　（源代码：array2.coffee）

```
[ 'a', 'b', 'c' ]
```

尽管上面的例子中这种定义数组的方式会让代码变得比较长，但你会发现，把数组的定义拆分成多行会让代码可读性更强。除此之外，还可以将换行和逗号结合起来一起使用，来增强代码的可读性。

例　（源代码：array3.coffee）

```
myArray = [
            "a", "b", "c"
```

```
            "d", "e", "f"
            "g", "h", "i"
          ]

console.log myArray
```

例 （源代码：array3.js）

```
(function() {
  var myArray;

  myArray = ["a", "b", "c", "d", "e", "f", "g", "h", "i"];

  console.log(myArray);

}).call(this);
```

输出 （源代码：array3.coffee）

```
[ 'a', 'b', 'c', 'd', 'e', 'f', 'g', 'h', 'i' ]
```

5.1.1　检测是否包含

JavaScript 中与数组相关的函数在 CoffeeScript 中也依然可用。而在 JavaScript 中相对比较困难的对数组的操作，在 CoffeeScript 中会变得容易许多。具体在本章后续介绍区间（range）时再详述。现在，先来介绍一种在 CoffeeScript 中变得尤为容易的数组操作：判断数据中是否存在某个特定的值。

例 （源代码：in_array.coffee）

```
myArray = ["a", "b", "c"]

if "b" in myArray
  console.log "I found 'b'."

unless "d" in myArray
  console.log "'d' was nowhere to be found."
```

例 （源代码：in_array.js）

```
(function() {
  var myArray,
    __indexOf = Array.prototype.indexOf || function(item) { for (var i = 0, l =
➥this.length; i < l; i++) { if (i in this && this[i] === item) return i; }
➥return -1; };

  myArray = ["a", "b", "c"];

  if (__indexOf.call(myArray, "b") >= 0) console.log("I found 'b'.");
```

```
  if (__indexOf.call(myArray, "d") < 0) {
    console.log("'d' was nowhere to be found.");
  }

}).call(this);
```

输出 （源代码：in_array.coffee）

```
I found 'b'.
'd' was nowhere to be found.
```

如上述代码所示，使用 in 关键字就可以检测数组中是否存在某个特定的值。在编译后的 JavaScript 代码中，会创建一个函数，用于遍历整个数组，并查找该值在数组中的下标。如果找到，则返回该下标，否则，就返回–1。后续代码就可以根据返回的下标是否大于 0 来判断数组中是否包含该元素了。

5.1.2 交换赋值

在编程中，有时会需要交换几个变量的值。这在某些语言中恐怕得写上好几行代码才能实现，但是，这在 CoffeeScript 中却非常简单。

我们来看一个交换两个变量值的例子。

例 （源代码：swap_assignment.coffee）

```
x = "X"
y = "Y"

console.log "x is #{x}"
console.log "y is #{y}"

[x, y] = [y, x]

console.log "x is #{x}"
console.log "y is #{y}"
```

例 （源代码：swap_assignment.js）

```
(function() {
  var x, y, _ref;

  x = "X";

  y = "Y";

  console.log("x is " + x);
```

```
  console.log("y is " + y);

  _ref = [y, x], x = _ref[0], y = _ref[1];

  console.log("x is " + x);

  console.log("y is " + y);

}).call(this);
```

输出 （源代码：swap_assignment.coffee）

```
x is X
y is Y
x is Y
y is X
```

CoffeeScript 中使用数组的语法来创建两个数组，并将数组中的值对应着进行交换。第一个数组，即=左边的数组，放置要赋新值的变量；第二个数组，即=右边的数组，放置要赋给变量的新值。这样就可以了，接下来的事情交给 CoffeeScript 处理就好了。

通过交换前后两次的输出结果就能发现这两个变量的确交换了各自的值。

5.1.3 多重赋值（又称解构赋值）

有时，我们会有这样的需求：将一个函数返回的数组中的值，挨个快速赋值给稍后要访问的变量。这在 Ruby 中很常见。Ruby 中有个很流行的库叫 Rack[①]，Rack 函数提供了 web 服务器和应用框架之间的简单的调用接口。其标准也很简单。框架必须返回一个数组，数组第一个元素是 HTTP 状态码，第二个元素是 HTTP 头，第三个也是最后一个元素是响应体内容。事实上 Rack 肯定没有刚刚描述的这么简单，不过对于我们来说够了，不需要了解太多。

下面我们来写一个函数，返回一个满足 Rack 标准的数组。然后，将数组中的值赋给相应的变量。

例 （源代码：multiple_assignment.coffee）

```
rack = ->
  [200, {"Content-Type": "text/html"}, "Hello Rack!"]

console.log rack()

[status, headers, body] = rack()

console.log "status is #{status}"
console.log "headers is #{JSON.stringify(headers)}"
console.log "body is #{body}"
```

① http://rack.rubyforge.org/

例 (源代码: multiple_assignment.js)

```
(function() {
  var body, headers, rack, status, _ref;
  rack = function() {
    return [
     200, {
       "Content-Type": "text/html"
     }, "Hello Rack!"
   ];
  };

  console.log(rack());

  _ref = rack(), status = _ref[0], headers = _ref[1], body = _ref[2];

  console.log("status is " + status);

  console.log("headers is " + (JSON.stringify(headers)));

  console.log("body is " + body);

}).call(this);
```

输出 (源代码: multiple_assignment.coffee)

```
[ 200, { 'Content-Type': 'text/html' }, 'Hello Rack!' ]
status is 200
headers is {"Content-Type":"text/html"}
body is Hello Rack!
```

上述例子中，使用了和交换赋值同样的方式。=左侧的数组中包含了要赋值的变量；=右边是调用函数后返回的数组，数组中的值会赋值给左侧数组中的变量。

> **提示：**不论是交换赋值还是多重赋值，都不需要去声明要使用的变量。CoffeeScript 会处理这些事情。

在多重赋值时，甚至都可以使用第 4 章中介绍的 splat。

例 (源代码: splat_assignment.coffee)

```
myArray = ["A", "B", "C", "D"]

[start, middle..., end] = myArray

console.log "start is #{start}"
console.log "middle is #{middle}"
console.log "end is #{end}"
```

例（源代码：splat_assignment.js）

```
(function() {
  var end, middle, myArray, start, _i,
    __slice = Array.prototype.slice;

  myArray = ["A", "B", "C", "D"];

  start = myArray[0], middle = 3 <= myArray.length ? __slice.call(myArray, 1, _i =
➥myArray.length - 1) : (_i = 1, []), end = myArray[_i++];

  console.log("start is " + start);

  console.log("middle is " + middle);

  console.log("end is " + end);

}).call(this);
```

输出（源代码：splat_assignment.coffee）

```
start is A
middle is B,C
end is D
```

要是右侧值的个数少于左侧变量的个数会怎么样呢？我们来看一个例子。

例（源代码：too_much_assignment.coffee）

```
myArray = ["A", "B"]

[a, b, c] = myArray

console.log "a is #{a}"
console.log "b is #{b}"
console.log "c is #{c}"
```

例（源代码：too_much_assignment.js）

```
(function() {
  var a, b, c, myArray;

  myArray = ["A", "B"];

  a = myArray[0], b = myArray[1], c = myArray[2];

  console.log("a is " + a);

  console.log("b is " + b);

  console.log("c is " + c);

}).call(this);
```

输出 (源代码: too_much_assignment.coffee)

```
a is A
b is B
c is undefined
```

如上述例子所示，最后一个变量的值为 undefined，因为右侧没有对应的值赋值给它。如果左侧变量的个数少于右侧值的个数，则剩余的值就不会被赋值给变量，如下例所示。

例 (源代码: too_little_assignment.coffee)

```
myArray = ["A", "B", "C"]

[a, b] = myArray

console.log "a is #{a}"
console.log "b is #{b}"
```

例 (源代码: too_little_assignment.js)

```
(function() {
  var a, b, myArray;

  myArray = ["A", "B", "C"];

  a = myArray[0], b = myArray[1];

  console.log("a is " + a);

  console.log("b is " + b);

}).call(this);
```

输出 (源代码: too_little_assignment.coffee)

```
a is A
b is B
```

5.2 区间

CoffeeScript 中的区间 (range) 能让定义包含两个数字之间所有数字的数组变得很容易。构建区间的语法如下。

例 (源代码: range1.coffee)

```
myRange = [1..10]
console.log myRange
```

例（源代码：range1.js）

```
(function() {
  var myRange;

  myRange = [1, 2, 3, 4, 5, 6, 7, 8, 9, 10];

  console.log(myRange);

}).call(this);
```

输出（源代码：range1.coffee）

```
[ 1, 2, 3, 4, 5, 6, 7, 8, 9, 10 ]
```

通过在起始数值和结束数值之间使用..可以获得一个数组，其元素包括了起始数值、结束数值以及这两个数值之间的所有数值。如果不想包括结束数值的话，可以使用...来代替..。

例（源代码：range2.coffee）

```
myRange = [1...10]
console.log myRange
```

例（源代码：range2.js）

```
(function() {
  var myRange;

  myRange = [1, 2, 3, 4, 5, 6, 7, 8, 9];

  console.log(myRange);

}).call(this);
```

输出（源代码：range2.coffee）

```
[ 1, 2, 3, 4, 5, 6, 7, 8, 9 ]
```

如上述例子所示，使用...来代替..之后，数组中就不包括 10 了。

用区间还可以构建逆序的数组。

例（源代码：range3.coffee）

```
myRange = [10..1]
console.log myRange
```

例（源代码：range3.js）

```
(function() {
  var myRange;
```

```
  myRange = [10, 9, 8, 7, 6, 5, 4, 3, 2, 1];

  console.log(myRange);

}).call(this);
```

输出（源代码：range3.coffee）

```
[ 10, 9, 8, 7, 6, 5, 4, 3, 2, 1 ]
```

构建逆序数组时，..和...也都可以使用，规则与之前的一样。

如上述例子所示，CoffeeScript 会在编译后的 JavaScript 代码中创建一个数组，并将所有的数值挨个完整写在数组定义中。尽管这样很清楚也很简单，但是问题来了，如果要定义一个包含成百上千个数值的数组怎么办？难道 CoffeeScript 也要构建一个巨大的 JavaScript 块把括号中所有数值都列出来吗？当然不会，当 CoffeeScript 发现数组元素个数超过一定数量时，就会采用循环来创建该区间中的元素。

例（源代码：range4.coffee）

```
myRange = [1..50]
console.log myRange.join(", ")
```

例（源代码：range4.js）

```
(function() {
  var myRange, _i, _results;

  myRange = (function() {
    _results = [];
    for (_i = 1; _i <= 50; _i++){ _results.push(_i); }
    return _results;
  }).apply(this);

  console.log(myRange.join(", "));

}).call(this);
```

输出（源代码：range4.coffee）

```
1, 2, 3, 4, 5, 6, 7, 8, 9, 10, 11, 12, 13, 14, 15, 16, 17, 18, 19, 20, 21, 22,
23, 24, 25, 26, 27, 28, 29, 30, 31, 32, 33, 34, 35, 36, 37, 38, 39, 40, 41, 42,
43, 44, 45, 46, 47, 48, 49, 50
```

提示：如果你想知道在 CoffeeScript 中，区间中的数字个数超过多少时算大，代码无法填充，我可以告诉你是 22。我不知道为什么是 22，只知道它就是 22。如果你不信，可以自己试一下。我已经试过了。

5.2.1 分割数组

区间非常强大，利用它可以有多种方法轻松地实现分割数组。

例（源代码：slice_array1.coffee）

```
myArray = [1..10]

firstThree = myArray[0..2]
console.log firstThree
```

例（源代码：slice_array1.js）

```
(function() {
  var firstThree, myArray;

  myArray = [1, 2, 3, 4, 5, 6, 7, 8, 9, 10];

  firstThree = myArray.slice(0, 3);

  console.log(firstThree);

}).call(this);
```

输出（源代码：slice_array1.coffee）

```
[ 1, 2, 3 ]
```

当然，这里还可以使用 0...3，它等效于 0..2。

例（源代码：slice_array2.coffee）

```
myArray = [1..10]

firstThree = myArray[0...3]
console.log firstThree
```

例（源代码：slice_array2.js）

```
(function() {
  var firstThree, myArray;

  myArray = [1, 2, 3, 4, 5, 6, 7, 8, 9, 10];

  firstThree = myArray.slice(0, 3);

  console.log(firstThree);

}).call(this);
```

输出（源代码：slice_array2.coffee）

```
[ 1, 2, 3 ]
```

不必限制自己只获取数组的第一部分，通过分割数组可以获取到数组的任意部分。

例（源代码：slice_array3.coffee）

```
myArray = [1..10]

middle = myArray[4..7]
console.log middle
```

例（源代码：slice_array3.js）

```
(function() {
  var middle, myArray;

  myArray = [1, 2, 3, 4, 5, 6, 7, 8, 9, 10];

  middle = myArray.slice(4, 8);

  console.log(middle);

}).call(this);
```

输出（源代码：slice_array3.coffee）

```
[ 5, 6, 7, 8 ]
```

如上例所示，还可以获取数组的中间部分，很酷吧！

5.2.2　替换数组值

区间及其语法的强大之处远不止如此，还可以用区间语法来替换数组中一段区域的值。

例（源代码：replace_array.coffee）

```
myArray = [1..10]
console.log myArray

myArray[4..7] = ['a', 'b', 'c', 'd']
console.log myArray
```

例（源代码：replace_array.js）

```
(function() {
  var myArray, _ref;
```

```
  myArray = [1, 2, 3, 4, 5, 6, 7, 8, 9, 10];

  console.log(myArray);

  [].splice.apply(myArray, [4, 4].concat(_ref = ['a', 'b', 'c', 'd'])), _ref;

  console.log(myArray);

}).call(this);
```

输出（源代码：replace_array.coffee）

```
[ 1, 2, 3, 4, 5, 6, 7, 8, 9, 10 ]
[ 1, 2, 3, 4, 'a', 'b', 'c', 'd', 9, 10 ]
```

我敢打赌你没见过这样的！确实很强大。

5.2.3　注入数值

有时，会需求要将一个数组中的值注入另一个数组的特定位置。为了实现该需求，我们使用一种和替换值的区间类似的技术。唯一的不同是不定义区间的结束数值，我们使用-1。

例（源代码：injecting_values.coffee）

```
myArray = [1..10]
console.log myArray

myArray[4..-1] = ['a', 'b', 'c', 'd']
console.log myArray
```

例（源代码：injecting_values.js）

```
(function() {
  var myArray, _ref;

  myArray = [1, 2, 3, 4, 5, 6, 7, 8, 9, 10];

  console.log(myArray);

  [].splice.apply(myArray, [4, -1 - 4 + 1].concat(_ref = ['a', 'b', 'c', 'd'])), _ref;

  console.log(myArray);

}).call(this);
```

输出（源代码：injecting_values.coffee）

```
[ 1, 2, 3, 4, 5, 6, 7, 8, 9, 10 ]
[ 1, 2, 3, 4, 'a', 'b', 'c', 'd', 5, 6, 7, 8, 9, 10 ]
```

如上述例子所示，我们将第二个数组中的元素注入了第一个数组中第 5 个元素开始的位置。原本该位置的元素为了给注入的新元素腾出空间会相应地往后移动。

5.3　对象/散列

JavaScript 的对象非常简单，基本就是包含了键/值对形式信息的数据结构。对象的值可以是其他对象、函数、数值或者字符串，等等。

> **提示**：JavaScript 和 CoffeeScript 中调用的对象其实就是典型的哈希表、哈希映射或者就是简单的散列。之所以冠以"对象"之名，我个人觉得多少有点用词不当，因为 JavaScript 中有其他类型的对象。我倾向于把它们看做是键/值对的集合[①]。

可以创建 CoffeeScript 中最基本的对象，方法如下。

例（源代码：basic_object.coffee）

```
obj = {}

console.log obj
```

例（源代码：basic_object.js）

```
(function() {
  var obj;

  obj = {};

  console.log(obj);

}).call(this);
```

输出（源代码：basic_object.coffee）

```
{}
```

不得不承认，这也太简单了，太过乏味了。下面我们来点"调料"，给对象加上一些键/值对数据。

例（源代码：basic_object2.coffee）

```
obj =
  firstName: "Mark"
  lastName: "Bates"

console.log obj
```

① 个人感觉这里作者的说法多少有点儿错误，这个问题和 JavaScript 中的类型有关，要搞清楚这块内容，个人推荐阅读周爱民老师的文章 http://blog.csdn.net/aimingoo/article/details/6676530。——译者注

例（源代码：basic_object2.js）

```
(function() {
  var obj;

  obj = {
    firstName: "Mark",
    lastName: "Bates"
  };

  console.log(obj);

}).call(this);
```

输出（源代码：basic_object2.coffee）

```
{ firstName: 'Mark', lastName: 'Bates' }
```

注意，因为键/值是分散在多行列出来的，所以无需再用逗号来分隔键/值对。另外，还可以去掉键/值对外面的大括号，因为这里定义对象时用了多行语法。我们也可以将定义对象写在一行上，如下所示。

例（源代码：basic_object2_single.coffee）

```
obj = { firstName: "Mark", lastName: "Bates" }

console.log obj
```

例（源代码：basic_object2_single.js）

```
(function() {
  var obj;

  obj = {
    firstName: "Mark",
    lastName: "Bates"
  };

  console.log(obj);

}).call(this);
```

输出（源代码：basic_object2_single.coffee）

```
{ firstName: 'Mark', lastName: 'Bates' }
```

尽管写在一行减少了代码行数，但不得不用到大括号和逗号，降低了可读性。

要想将函数添加到对象中也很容易。

例 （源代码：basic_object3.coffee）

```
obj =
  firstName: "Mark"
  lastName: "Bates"
  fullName: ->
    "#{@firstName} #{@lastName}"

  console.log obj
```

例 （源代码：basic_object3.js）

```
(function() {
  var obj;

  obj = {
    firstName: "Mark",
    lastName: "Bates",
    fullName: function() {
      return "" + this.firstName + " " + this.lastName;
    }
  };

  console.log(obj);

}).call(this);
```

输出 （源代码：basic_object3.coffee）

```
{ firstName: 'Mark', lastName: 'Bates', fullName: [Function] }
```

在创建新对象时，CoffeeScript 中还有一个小技巧。我们有时在创建对象时想要用一些变量作为值，同时又想让对象中的键名和这些变量名一样，就像下面那样。

例 （源代码：object_keys1.coffee）

```
foo = 'FOO'
bar = 'BAR'

obj =
  foo: foo
  bar: bar

console.log obj
```

例 （源代码：object_keys1.js）

```
(function() {
  var bar, foo, obj;
```

```
  foo = 'FOO';

  bar = 'BAR';

  obj = {
    foo: foo,
    bar: bar
  };

  console.log(obj);

}).call(this);
```

输出（源代码: object_keys1.coffee）

```
{ foo: 'FOO', bar: 'BAR' }
```

你是不是觉得这种写法有点多此一举？其实我也这么觉得。幸运的是，CoffeeScript 也有同感。在 CoffeeScript 中，我们可以用如下这种方式来简化此前的写法。

例（源代码: object_keys2.coffee）

```
foo = 'FOO'
bar = 'BAR'

obj = {
  foo
  bar
}

console.log obj
```

例（源代码: object_keys2.js）

```
(function() {
  var bar, foo, obj;

  foo = 'FOO';

  bar = 'BAR';

  obj = {
    foo: foo,
    bar: bar
  };

  console.log(obj);

}).call(this);
```

输出（源代码：object_keys2.coffee）

```
{ foo: 'FOO', bar: 'BAR' }
```

使用这种简化方式的代价就是必须要加上大括号；否则，CoffeeScript 就无法正确解析了。

最后，如果定义一个对象作为函数调用的一部分，大括号可以不加，用多行或者单行方式来代替。

例（源代码：objects_into_functions.coffee）

```
myFunc = (options) ->
  console.log options

myFunc(foo: 'Foo', bar: 'Bar')
```

例（源代码：objects_into_functions.js）

```
(function() {
  var myFunc;

  myFunc = function(options) {
    return console.log(options);
  };

  myFunc({
    foo: 'Foo',
    bar: 'Bar'
  });

}).call(this);
```

输出（源代码：objects_into_functions.coffee）

```
{ foo: 'Foo', bar: 'Bar' }
```

好了，女士们、先生们，以上就是所有 CoffeeScript 中定义对象的方式了。

5.3.1　设置属性/获取属性

使用对象时，绝大部分情况下都会希望能够在代码中获取存储在对象中的值。在 CoffeeScript 中获取这些值的方式和在 JavaScript 中没什么区别。可以使用 `.` 或者使用 `[]` 来进行访问。

例（源代码：object_get_attributes.coffee）

```
obj =
  firstName: "Mark"
```

```
  lastName: "Bates"
  fullName: ->
    "#{@firstName} #{@lastName}"

console.log obj.firstName
console.log obj['lastName']
console.log obj.fullName()
```

例（源代码：object_get_attributes.js）

```
(function() {
  var obj;

  obj = {
    firstName: "Mark",
    lastName: "Bates",
    fullName: function() {
      return "" + this.firstName + " " + this.lastName;
    }
  };

  console.log(obj.firstName);

  console.log(obj['lastName']);

  console.log(obj.fullName());

}).call(this);
```

输出（源代码：object_get_attributes.coffee）

```
Mark
Bates
Mark Bates
```

设置属性的方式也是如此。

例（源代码：object_set_attributes.coffee）

```
obj =
  firstName: "Mark"
  lastName: "Bates"
  fullName: ->
    "#{@firstName} #{@lastName}"

obj.firstName = 'MARK'
console.log obj.firstName
obj['lastName'] = 'BATES'
console.log obj['lastName']
```

例（源代码：object_set_attributes.js）

```
(function() {
  var obj;

  obj = {
    firstName: "Mark",
    lastName: "Bates",
    fullName: function() {
      return "" + this.firstName + " " + this.lastName;
    }
  };

  obj.firstName = 'MARK';

  console.log(obj.firstName);

  obj['lastName'] = 'BATES';

  console.log(obj['lastName']);

}).call(this);
```

输出（源代码：object_set_attributes.coffee）

```
MARK
BATES
```

> **提示**：流行的 JavaScript 校验框架——JSLint[①]，推荐使用.来获取对象中的属性。我表示认同。我发现用.读起来感觉更好，并且可以减少代码量，非常好。

5.3.2　解构赋值

前面介绍数组时介绍过，CoffeeScript 中可以将数组中特定一些元素直接赋值给变量。同样，在 CoffeeScript 中对象也可以这样。

从对象中解构出值的语法和解构数组的方式不尽相同。其语法更像是定义对象的语法，不同的只是定义对象时用的是键/值对，而解构是列出键值。

下面举例来说可能会清楚一点。

例（源代码：object_destructuring.coffee）

```
book =
  title: "Distributed Programming with Ruby"
```

① http://www.jslint.com/

```
  author: "Mark Bates"
  chapter_1:
    name: "Distributed Ruby (DRb)"
    pageCount: 33
  chapter_2:
    name: "Rinda"
    pageCount: 40

{author, chapter_1: {name, pageCount}} = book

console.log "Author: #{author}"
console.log "Chapter 1: #{name}"
console.log "Page Count: #{pageCount}"
```

例（源代码: object_destructuring.js）

```
(function() {
  var author, book, name, pageCount, _ref;

  book = {
    title: "Distributed Programming with Ruby",
    author: "Mark Bates",
    chapter_1: {
      name: "Distributed Ruby (DRb)",
      pageCount: 33
    },
    chapter_2: {
      name: "Rinda",
      pageCount: 40
    }
  };

  author = book.author, (_ref = book.chapter_1, name = _ref.name, pageCount =
➥_ref.pageCount);

  console.log("Author: " + author);

  console.log("Chapter 1: " + name);

  console.log("Page Count: " + pageCount);

}).call(this);
```

输出（源代码: object_destructuring.coffee）

```
Author: Mark Bates
Chapter 1: Distributed Ruby (DRb)
Page Count: 33
```

5.4 循环与迭代

对于绝大多数应用而言，都必须要能够迭代对象的键、值以及数组。我们可能想要在屏幕上打印出一个书单，或者要修改对象中所有的值。不论什么样的应用，在 CoffeeScript 中迭代它们都是非常简单的。我们来看看吧。

5.4.1 迭代数组

在 JavaScript 中，我最不怎么喜欢做的就是迭代数组。因为既麻烦又容易出问题。CoffeeScript 则实现了一个类似 Ruby 语言中的 `for` 循环结构来进行迭代。

循环结构非常简单：

```
for <some name here> in <array here>
```

定义好 `for` 循环之后，通过缩进来定义每次迭代要执行的代码，就像第 3 章中的 `if` 语句和 `else` 语句那样。

下面这个例子，我们对一个包含字母的数组进行循环，并将它们以大写形式打印出来。

例（源代码：iterating_arrays.coffee）

```
myLetters = ["a", "b", "c", "d"]

for letter in myLetters
  console.log letter.toUpperCase()
```

例（源代码：iterating_arrays.js）

```
(function() {
  var letter, myLetters, _i, _len;

  myLetters = ["a", "b", "c", "d"];

  for (_i = 0, _len = myLetters.length; _i < _len; _i++) {
    letter = myLetters[_i];
    console.log(letter.toUpperCase());
  }

}).call(this);
```

输出（源代码：iterating_arrays.coffee）

```
A
B
C
D
```

1. by 关键字

假设我们有一个包含 26 个英文字母的数组，现在我们要将这些字母每隔一个打印出来。要实现此需求，可以在定义 for 循环时使用 by 关键字：

例（源代码：iterating_arrays_by.coffee）

```
letters = ["a", "b", "c", "d", "e", "f", "g", "h", "i", "j", "k", "l", "m", "n", "o",
➥"p", "q", "r", "s", "t", "u", "v", "w", "x", "y", "z"]
for letter in letters by 2
  console.log letter
```

例（源代码：iterating_arrays_by.js）

```
(function() {
  var letter, letters, _i, _len, _step;

  letters = ["a", "b", "c", "d", "e", "f", "g", "h", "i", "j", "k", "l", "m", "n",
➥"o", "p", "q", "r", "s", "t", "u", "v", "w", "x", "y", "z"];

  for (_i = 0, _len = letters.length, _step = 2; _i < _len; _i += _step) {
    letter = letters[_i];
    console.log(letter);
  }

}).call(this);
```

输出（源代码：iterating_arrays_by.coffee）

```
a
c
e
g
i
k
m
o
q
s
u
w
y
```

by 关键字后可以是任意数字，for 循环会相应地对数组做迭代。

2. when 关键字

when 关键字可以给 for 循环添加一个简单的条件。

假设有一个包含从 1 到 10 十个数字的数组，但我们只想将小于 5 的数打印出来。这时，我们就可以将代码写成如下形式。

例（源代码：iterating_with_when1.coffee）

```
a = [1..10]

for num in a
  if num < 5
    console.log num
```

例（源代码：iterating_with_when1.js）

```
(function() {
  var a, num, _i, _len;

  a = [1, 2, 3, 4, 5, 6, 7, 8, 9, 10];

  for (_i = 0, _len = a.length; _i < _len; _i++) {
    num = a[_i];
    if (num < 5) console.log(num);
  }

}).call(this);
```

输出（源代码：iterating_with_when1.coffee）

```
1
2
3
4
```

我们也可以在 for 循环定义之后使用 when 关键字来改写上述例子。

例（源代码：iterating_with_when2.coffee）

```
a = [1..10]

for num in a when num < 5
  console.log num
```

例（源代码：iterating_with_when2.js）

```
(function() {
  var a, num, _i, _len;

  a = [1, 2, 3, 4, 5, 6, 7, 8, 9, 10];

  for (_i = 0, _len = a.length; _i < _len; _i++) {
    num = a[_i];
    if (num < 5) console.log(num);
  }

}).call(this);
```

输出 （源代码：iterating_with_when2.coffee）

```
1
2
3
4
```

5.4.2 迭代对象

在 CoffeeScript 中，迭代对象和迭代数组的方式差不多。

迭代对象的 for 循环语法如下：

```
for <key name here>, <value name here> of <object here>
```

我们来看一个例子。

例 （源代码：iterating_objects.coffee）

```
person =
  firstName: "Mark"
  lastName: "Bates"

for key, value of person
  console.log "#{key} is #{value}"
```

例 （源代码：iterating_objects.js）

```
(function() {
  var key, person, value;

  person = {
    firstName: "Mark",
    lastName: "Bates"
  };

  for (key in person) {
    value = person[key];
    console.log("" + key + " is " + value);
  }

}).call(this);
```

输出 （源代码：iterating_objects.coffee）

```
firstName is Mark
lastName is Bates
```

迭代对象与迭代数组在 for 循环语法上有两大明显区别：第一，迭代对象时，我们需要在 for 循环中为对象定义两个变量，一个代表键、一个代表值，分别对应对象中的键/值对。另外一个不同就是迭代对象时使用的关键字是 of，而不是迭代数组时使用的 in。

1. by 关键字

很遗憾，迭代对象时不能使用 by 关键字，原因是在对象中不可能像数组那样实现一个隔着一个地访问键值对。

2. when 关键字

与 by 关键字不同，定义对象的 for 循环时可以使用 when 关键字。

下面这个例子，我们只把值长度小于 5 的键/值对打印出来。

例（源代码：iterating_objects_with_when.coffee）

```
person =
  firstName: "Mark"
  lastName: "Bates"

for key, value of person when value.length < 5
  console.log "#{key} is #{value}"
```

例（源代码：iterating_objects_with_when.js）

```
(function() {
  var key, person, value;

  person = {
    firstName: "Mark",
    lastName: "Bates"
  };

  for (key in person) {
    value = person[key];
    if (value.length < 5) console.log("" + key + " is " + value);
  }

}).call(this);
```

输出（源代码：iterating_objects_with_when.coffee）

```
firstName is Mark
```

3. own 关键字

在 JavaScript 中，可以使用 prototype[①]函数给所有系统对象添加函数或者属性值。像

① http://en.wikipedia.org/wiki/JavaScript#Prototype-based

jQuery 这样的库就是使用这种方式给数组、字符串等对象添加了特殊函数的。

下面来看一个例子。

例 （源代码：iterating_objects_without_own.coffee）

```
myObject =
  name: "Mark"

for key, value of myObject
  console.log "#{key}: #{value}"

Object.prototype.dob = new Date(1976, 7, 24)

for key, value of myObject
  console.log "#{key}: #{value}"

anotherObject =
  name: "Bob"

for key, value of anotherObject
  console.log "#{key}: #{value}"
```

例 （源代码：iterating_objects_without_own.js）

```
(function() {
  var anotherObject, key, myObject, value;

  myObject = {
    name: "Mark"
  };

  for (key in myObject) {
    value = myObject[key];
    console.log("" + key + ": " + value);
  }

  Object.prototype.dob = new Date(1976, 7, 24);

  for (key in myObject) {
    value = myObject[key];
    console.log("" + key + ": " + value);
  }

  anotherObject = {
    name: "Bob"
  };

  for (key in anotherObject) {
    value = anotherObject[key];
    console.log("" + key + ": " + value);
```

```
  }

}).call(this);
```

输出（源代码：iterating_objects_without_own.coffee）

```
name: Mark
name: Mark
dob: Tue Aug 24 1976 00:00:00 GMT-0400 (EDT)
name: Bob
dob: Tue Aug 24 1976 00:00:00 GMT-0400 (EDT)
```

在第一次循环 myObject 中的键/值对时，只有我们定义的 name 值。而给 object 的原型（prototype）加上 dob 之后，再次循环 myObject 对象中的键/值对时，我们看到 dob 的值也打印出来了。

那我们怎么样才能只看到对象中我们显式定义的键/值对呢？在 JavaScript 中，可以使用 hasOwnProperty 函数来检测键是显式定义的（即直接由对象定义的）还是继承自 object 原型上的。而在 CoffeeScript 中，只需将 for 循环换成 for own 循环即可。

例（源代码：iterating_objects_with_own.coffee）

```
myObject =
  name: "Mark"

for own key, value of myObject
  console.log "#{key}: #{value}"

Object.prototype.dob = new Date(1976, 7, 24)

for own key, value of myObject
  console.log "#{key}: #{value}"

anotherObject =
  name: "Bob"

for own key, value of anotherObject
  console.log "#{key}: #{value}"
```

例（源代码：iterating_objects_with_own.js）

```
(function() {
  var anotherObject, key, myObject, value,
    __hasProp = Object.prototype.hasOwnProperty;


  myObject = {
    name: "Mark"
  };
```

```
  for (key in myObject) {
    if (!__hasProp.call(myObject, key)) continue;
    value = myObject[key];
    console.log("" + key + ": " + value);
  }

  Object.prototype.dob = new Date(1976, 7, 24);

  for (key in myObject) {
    if (!__hasProp.call(myObject, key)) continue;
    value = myObject[key];
    console.log("" + key + ": " + value);
  }

  anotherObject = {
    name: "Bob"
  };

  for (key in anotherObject) {
    if (!__hasProp.call(anotherObject, key)) continue;
    value = anotherObject[key];
    console.log("" + key + ": " + value);
  }

}).call(this);
```

输出 （源代码：iterating_objects_with_own.coffee）

```
name: Mark
name: Mark
name: Bob
```

完美！现在我们只获取了定义在 myObject 上的键/值对。

5.4.3　while 循环

作为开发者，有时我们会有这样的需求：当特定条件为 true 时，需要重复执行一段代码。例如，重复打印 n 次信息，或者在加载文件时，显示“请稍等”之类的文字。要实现此类需求，可以使用 CoffeeScript 中的 while 循环。

我们来写一个将一段代码执行 n 次的函数。

例 （源代码：while_loop.coffee）

```
times = (number_of_times, callback)->
  index = 0
  while index++ < number_of_times
    callback(index)
```

```
  return null

times 5, (index)->
  console.log index
```

例 (源代码: while_loop.js)

```
(function() {
  var times;

  times = function(number_of_times, callback) {
    var index;
    index = 0;
    while (index++ < number_of_times) {
      callback(index);
    }
    return null;
  };

  times(5, function(index) {
    return console.log(index);
  });

}).call(this);
```

输出 (源代码: while_loop.coffee)

```
1
2
3
4
5
```

上述例子中，在times函数里有一个while循环，在index小于传入的number_of_times参数时，始终去执行callback函数。

> **提示**：在while循环的例子中，我们看到了index++这样的代码。可能有人对++操作符还不熟悉，其实它的作用就是将变量值加1之后得到的新值再赋给该变量，等效于index = index + 1。

5.4.4 until 循环

顾名思义，until循环和while循环是相对的。while循环是在条件为true时始终执行一段代码，until循环则是在条件为false时始终执行一段代码。

我们可以用until循环来改写此前while循环的例子。

例（源代码：until_loop.coffee）

```
times = (number_of_times, callback)->
  index = 0
  until index++ >= number_of_times
    callback(index)
  return null

times 5, (index)->
  console.log index
```

例（源代码：until_loop.js）

```
(function() {
  var times;

  times = function(number_of_times, callback) {
    var index;
    index = 0;
    while (!(index++ >= number_of_times)) {
      callback(index);
    }
    return null;
  };

  times(5, function(index) {
    return console.log(index);
  });

}).call(this);
```

输出（源代码：until_loop.coffee）

```
1
2
3
4
5
```

> **提示**：下面这句话有助于更好地记清楚这几个循环：while 循环是当条件为 true 时才运行，until 循环一直运行到条件为 true 为止。这可能对你没什么用，不过确实有很多人搞不清楚这些。

5.5　comprehension

在很多迭代的例子中，要执行的代码段都非常简单，就像下面这样的。

例 （源代码：iterating_arrays.coffee）

```
myLetters = ["a", "b", "c", "d"]

for letter in myLetters
  console.log letter.toUpperCase()
```

因为在上面的 for 循环中用的是单行代码块，所以我们能用 CoffeeScript 中的 comprehension 语法糖了。comprehension 允许将循环及其代码块写在一行上。

下面这个例子展示了如何使用 comprehension。

例 （源代码：iterating_arrays_comprehension.coffee）

```
myLetters = ["a", "b", "c", "d"]

console.log letter.toUpperCase() for letter in myLetters
```

例 （源代码：iterating_arrays_comprehension.js）

```
(function() {
  var letter, myLetters, _i, _len;

  myLetters = ["a", "b", "c", "d"];

  for (_i = 0, _len = myLetters.length; _i < _len; _i++) {
    letter = myLetters[_i];
    console.log(letter.toUpperCase());
  }

}).call(this);
```

输出 （源代码：iterating_arrays_comprehension.coffee）

```
A
B
C
D
```

如上述代码所示，我们将代码块放到了 for 循环语句之前，写在了一行上。

还可以使用 comprehension 捕获 for 循环的结果。继续上面的例子，这次我们来捕获 for 循环的结果，将这些大写字母放到一个新数组中。

例 （源代码：iterating_arrays_comprehension_capture.coffee）

```
myLetters = ["a", "b", "c", "d"]

upLetters = (letter.toUpperCase() for letter in myLetters)

console.log upLetters
```

例（源代码：iterating_arrays_comprehension_capture.js）

```
(function() {
  var letter, myLetters, upLetters;

  myLetters = ["a", "b", "c", "d"];

  upLetters = (function() {
    var _i, _len, _results;
    _results = [];
    for (_i = 0, _len = myLetters.length; _i < _len; _i++) {
      letter = myLetters[_i];
      _results.push(letter.toUpperCase());
    }
    return _results;
  })();

  console.log(upLetters);

}).call(this);
```

输出（源代码：iterating_arrays_comprehension_capture.coffee）

```
[ 'A', 'B', 'C', 'D' ]
```

通过将 comprehension 语句包在括号中，我们可以将捕获的迭代结果赋值给另外一个变量。另外，值得一提的是，对于多行的 for 循环，也可以捕获其结果。

例（源代码：iterating_arrays_capture.coffee）

```
myLetters = ["a", "b", "c", "d"]

upLetters = for letter in myLetters
  letter.toUpperCase()

console.log upLetters
```

例（源代码：iterating_arrays_capture.js）

```
(function() {
  var letter, myLetters, upLetters;

  myLetters = ["a", "b", "c", "d"];

  upLetters = (function() {
    var _i, _len, _results;
    _results = [];
    for (_i = 0, _len = myLetters.length; _i < _len; _i++) {
      letter = myLetters[_i];
      _results.push(letter.toUpperCase());
```

```
    }
    return _results;
  })();

  console.log(upLetters);

}).call(this);
```

输出（源代码：iterating_arrays_capture.coffee）

```
[ 'A', 'B', 'C', 'D' ]
```

> **提示**：这部分介绍的内容我不提倡大家一定要这么用。CoffeeScript 强调了 comprehension 语法糖的“强大”。我也表示同意，不过，我个人觉得这样写出来的代码可读性会变差，并且维护起来也会比较困难。所以，使用它之前要先想清楚是否真的适合。

5.6 do 关键字

我们第 2 章中介绍的作用域是 JavaScript 中令人头疼的概念，特别是循环中的作用域问题 。由于 JavaScript 异步的天性，当我们做诸如循环访问几个数字这样的简单操作时，有可能会丢失变量的作用域。

我们来看下面这个例子。循环访问几个数字，并将它们打印出来。不过，在打印之前，我们想等 1 秒钟后再打印。

例（源代码：do.coffee）

```
for x in [1..5]
  setTimeout ->
    console.log x
  ,1
```

例（源代码：do.js）

```
(function() {
  var x;

  for (x = 1; x <= 5; x++) {
    setTimeout(function() {
      return console.log(x);
    }, 1);
  }

}).call(this);
```

输出（源代码：do.coffee）

```
6
6
6
6
6
```

这显然不是我们想要的！我们想要的是依次打印出 1 到 5 这 5 个数字。结果却是将 6 连续打印了 5 次，怎么回事儿？答案就是我们丢失了变量 x 所在的作用域。

在等待打印的过程中，变量随着循环的进行不断地在递增。之所以输出的结果是 6，原因就在于最后一次当递增到 6 时，正好大于 5，于是就跳出循环了。

那么如何才能避免这类事情发生，能够像我们预期的那样正确获取该变量值呢？这个时候可以使用 do 关键字。

例（源代码：do2.coffee）

```
for x in [1..5]
  do (x) ->
    setTimeout ->
      console.log x
    , 1
```

例（源代码：do2.js）

```
(function() {
  var x, _fn;

  _fn = function(x) {
    return setTimeout(function() {
      return console.log(x);
    }, 1);
  };
  for (x = 1; x <= 5; x++) {
    _fn(x);
  }

}).call(this);
```

输出（源代码：do2.coffee）

```
1
2
3
4
5
```

do 关键字会在要执行的代码外部创建一个包装函数，这样就能将要访问的代码捕获住。真的很实用！

5.7　小结

好了，至此，所有 CoffeeScript 中与集合与迭代相关的都介绍完了。

首先，我们介绍了数组以及 CoffeeScript 是如何处理数组的。我们还介绍了 CoffeeScript 中与数组相关的一些有趣的技巧：测试数组中是否包含某个值、变量之间的交换赋值，最后介绍了捕获数组元素并将其赋值给其他变量。

随后，我们介绍了区间。介绍了如何使用区间的语法来构造由数值组成的数组。还介绍了如何使用区间来操作现有的数组：截取数组的一部分以及用其他的值来替换数组中部分元素。

介绍完区间，我们又介绍了 CoffeeScript 中的对象，介绍了构造对象的不同规则，介绍了如何设置和获取我们创建的对象的属性。以及使用修改过的对象语法，如何获取对象中深度嵌套的属性并将它们赋值给其他变量。

紧接着又介绍了迭代。介绍了如何迭代数组中的元素以及如何迭代对象中的键/值对信息。还介绍了 `by` 关键字和 `when` 关键字来帮助我们编写更整洁的循环代码。之后，我们通过比较 `while` 循环和 `until` 循环，介绍了两者的区别。

我们还介绍了 comprehension 语法，有了它，我们可以将循环代码和循环内要执行的代码块写在一行上。

最后，我们介绍了 `do` 关键字，介绍了在对集合进行循环这类操作时，它是如何帮助捕获要访问的变量的作用域的。

有了集合的知识，我们可以继续介绍另一类集合——类。

第 6 章

类

类[1]对于用预先定义好的函数和变量来创建一个对象的实例是至关重要的。这些实例可以存储和该实例相关的状态。多年来，JavaScript 一直都因为不真正支持类而饱受攻击。

第 5 章介绍了 JavaScript 中的对象。其中每个例子，我们都给创建的新对象设置了一系列的函数和值。对于简单对象而言，这样做完全没有问题，不过，遇到更复杂的数据模型时，该怎么办呢？更重要的是，遇到多个这类复杂数据模型时，又该如何是好呢？类就是很好的解决方案。

幸运的是，CoffeeScript 全面支持类。你可能会问：既然 JavaScript 原生没有提供类的支持，CoffeeScript 又是如何解决这个问题的呢？简单来说，就是通过使用作用域和函数以及对象来解决。要知道更为详细的答案，请阅读本章后续内容。

6.1 定义类

在 CoffeeScript 中，定义类非常简单，只需一行代码即可。

例（源代码：simple_class1.coffee）

```
class Employee
```

例（源代码：simple_class1.js）

```
(function() {
  var Employee;

  Employee = (function() {

    function Employee() {}
```

① http://en.wikipedia.org/wiki/Class_(computer_programming)

```
    return Employee;

  })();

}).call(this);
```

在上述例子中，我们定义了一个简单的新的名为 Employee 的类。现在我们可以用如下方式来创建该类的实例了。

例（源代码：simple_class2.coffee）

```
class Employee

emp1 = new Employee()
emp2 = new Employee()
```

例（源代码：simple_class2.js）

```
(function() {
  var Employee, emp1, emp2;

  Employee = (function() {

    function Employee() {}

    return Employee;

  })();

  emp1 = new Employee();

  emp2 = new Employee();

}).call(this);
```

通过在类名前使用 new 关键字，就创建了该类的一个全新的实例，随后，就可以操作该实例了。

> **提示：** 在用 new 关键字创建对象的新实例时，最后的括号不是必须的，不过我觉得加上括号看上去会更清楚，可读性也更好。你可以根据自己的喜好来选择。

6.2 定义函数

在类中定义函数遵循的规则和语法和在简单对象中定义函数是一样的。

例（源代码：simple_class_with_function.coffee）

```
class Employee

  dob: (year = 1976, month = 7, day = 24)->
```

```
  new Date(year, month, day)

emp1 = new Employee()
console.log emp1.dob()
emp2 = new Employee()
console.log emp2.dob(1979, 3, 28)
```

例 （源代码: simple_class_with_function.js）

```
(function() {
  var Employee, emp1, emp2;

  Employee = (function() {

    function Employee() {}

    Employee.prototype.dob = function(year, month, day) {
      if (year == null) year = 1976;
      if (month == null) month = 7;
      if (day == null) day = 24;
      return new Date(year, month, day);
    };

    return Employee;

  })();

  emp1 = new Employee();

  console.log(emp1.dob());

  emp2 = new Employee();

  console.log(emp2.dob(1979, 3, 28));

}).call(this);
```

输出 （源代码: simple_class_with_function.coffee）

```
Tue, 24 Aug 1976 04:00:00 GMT
Sat, 28 Apr 1979 05:00:00 GMT
```

6.3 constructor 函数

CoffeeScript 允许定义称为构造器（constructor）的函数（构造函数），该函数会在创建实例的时候被调用。构造器和其他函数类似，唯一的不同就是，构造函数会在使用 new 关键字创建实例时被自动调用，不需要显式地去调用它。

提示：刚刚我说构造函数会在创建实例时被自动调用，这种说法不算很准确。事实上，像new Employee()这样创建实例时，其实也可以理解为是直接调用了该构造函数。

例（源代码：simple_class3.coffee）

```
class Employee

  constructor: ->
    console.log "Instantiated a new Employee object"

  dob: (year = 1976, month = 7, day = 24)->
    new Date(year, month, day)

emp1 = new Employee()
console.log emp1.dob()

emp2 = new Employee()
console.log emp2.dob(1979, 3, 28)
```

例（源代码：simple_class3.js）

```
(function() {
  var Employee, emp1, emp2;

  Employee = (function() {

    function Employee() {
      console.log("Instantiated a new Employee object");
    }

    Employee.prototype.dob = function(year, month, day) {
      if (year == null) year = 1976;
      if (month == null) month = 7;
      if (day == null) day = 24;
      return new Date(year, month, day);
    };

    return Employee;

  })();

  emp1 = new Employee();

  console.log(emp1.dob());

  emp2 = new Employee();

  console.log(emp2.dob(1979, 3, 28));

}).call(this);
```

输出 （源代码：simple_class3.coffee）

```
Instantiated a new Employee object
Tue, 24 Aug 1976 04:00:00 GMT
Instantiated a new Employee object
Sat, 28 Apr 1979 05:00:00 GMT
```

如上述例所示，每次我们创建新的 Employee 对象时，就会在控制台输出一段信息以此提示对象已经创建了。

本章后续内容会介绍，构造函数提供了一个快速为新对象设置自定义数据的简单方式。

6.4 类中的作用域

CoffeeScript 中的类本质上就是模拟真实世界的对象，定义了一大堆 JavaScript 样板代码，供实例化的对象使用。因为类就如一般对象——有一定板式的对象，所以其中变量、属性以及函数的作用域和它们在常规对象中都是一样的。

我们再来看看 Employee 类。Employee 在真实世界中有名字，所以在 Employee 类中也要体现出来。在创建一个新的 Employee 实例时，我们想要将名字传进去，将其赋值给一个属性，其作用域限定在该新的实例中。

例 （源代码：class_scope.coffee）

```
class Employee

  constructor: (name)->
    @name = name

  dob: (year = 1976, month = 7, day = 24)->
    new Date(year, month, day)

  emp1 = new Employee("Mark")
  console.log emp1.name
  console.log emp1.dob()

  emp2 = new Employee("Rachel")
  console.log emp2.name
  console.log emp2.dob(1979, 3, 28)
```

例 （源代码：class_scope.js）

```
(function() {
  var Employee, emp1, emp2;

  Employee = (function() {

    function Employee(name) {
```

```
    this.name = name;
  }

  Employee.prototype.dob = function(year, month, day) {
    if (year == null) year = 1976;
    if (month == null) month = 7;
    if (day == null) day = 24;
    return new Date(year, month, day);
  };

  return Employee;

 })();

 emp1 = new Employee("Mark");

 console.log(emp1.name);

 console.log(emp1.dob());

 emp2 = new Employee("Rachel");

 console.log(emp2.name);

 console.log(emp2.dob(1979, 3, 28));

}).call(this);
```

输出 （源代码：class_scope.coffee）

```
Mark
Tue, 24 Aug 1976 04:00:00 GMT
Rachel
Sat, 28 Apr 1979 05:00:00 GMT
```

之前我们介绍过，构造函数和其他函数没有什么区别，作用域和定义规则也是如此。上述例子中，我们将 name 参数传递进去，并用@name=name 将该参数赋值给对象实例的 name 属性。还记得第 3 章中介绍的吧，这里@就等于 this。

设置了该实例的属性之后，就可以像访问其他对象上的属性一样来访问实例上的属性了。

针对上述例子，还有一种更简便的方式，也是我个人最喜欢的 CoffeeScript 特性之一（希望其他语言也能实现这样的特性），我们可以将构造器简化为如下形式。

例 （源代码：class_scope1.coffee）

```
class Employee

  constructor: (@name)->

  dob: (year = 1976, month = 7, day = 24)->
    new Date(year, month, day)

emp1 = new Employee("Mark")
```

```
console.log emp1.name
console.log emp1.dob()

emp2 = new Employee("Rachel")
console.log emp2.name
console.log emp2.dob(1979, 3, 28)
```

例　（源代码：class_scope1.js）

```
(function() {
  var Employee, emp1, emp2;

  Employee = (function() {

    function Employee(name) {
      this.name = name;
    }

    Employee.prototype.dob = function(year, month, day) {
      if (year == null) year = 1976;
      if (month == null) month = 7;
      if (day == null) day = 24;
      return new Date(year, month, day);
    };

    return Employee;

  })();

  emp1 = new Employee("Mark");

  console.log(emp1.name);

  console.log(emp1.dob());

  emp2 = new Employee("Rachel");
  console.log(emp2.name);
  console.log(emp2.dob(1979, 3, 28));

}).call(this);
```

输出　（源代码：class_scope1.coffee）

```
Mark
Tue, 24 Aug 1976 04:00:00 GMT
Rachel
Sat, 28 Apr 1979 05:00:00 GMT
```

通过在 name 参数的定义前加上@操作符，告诉 CoffeeScript 要生成一段 JavaScript 代码来将该参数赋值给同名的属性，如上述例子中就是 name 属性。

从类中其他函数访问属性也很简单。我们来更新下上述例子，这次添加一个将员工名字和生日信息打印出来的函数。

例（源代码：class_scope2.coffee）

```
class Employee

  constructor: (@name)->

  dob: (year = 1976, month = 7, day = 24)->
    new Date(year, month, day)

  printInfo: (year = 1976, month = 7, day = 24)->
    console.log "Name: #{@name}"
    console.log "DOB: #{@dob(year, month, day)}"


emp1 = new Employee("Mark")
emp1.printInfo(1976, 7, 24)

emp2 = new Employee("Rachel")
emp2.printInfo(1979, 3, 28)
```

例（源代码：class_scope2.js）

```
(function() {
  var Employee, emp1, emp2;

  Employee = (function() {

    function Employee(name) {
      this.name = name;
    }

    Employee.prototype.dob = function(year, month, day) {
      if (year == null) year = 1976;
      if (month == null) month = 7;
      if (day == null) day = 24;
      return new Date(year, month, day);
    };

    Employee.prototype.printInfo = function(year, month, day) {
      if (year == null) year = 1976;
      if (month == null) month = 7;
      if (day == null) day = 24;
      console.log("Name: " + this.name);
      return console.log("DOB: " + (this.dob(year, month, day)));
    };

    return Employee;

  })();
```

```
  emp1 = new Employee("Mark");

  emp1.printInfo(1976, 7, 24);

  emp2 = new Employee("Rachel");

  emp2.printInfo(1979, 3, 28);

}).call(this);
```

输出 （源代码：class_scope2.coffee）

```
Name: Mark
DOB: Tue Aug 24 1976 00:00:00 GMT-0400 (EDT)
Name: Rachel
DOB: Sat Apr 28 1979 00:00:00 GMT-0500 (EST)
```

上述例子如果代码就写成这样了那我显然有点不负责任。我不想每次在打印出员工信息时，都要将年、月、日传递给 printInfo 函数。我想将生日的日期传递给构造函数，然后在需要的时候获取该日期。我们来将上述代码改写为如下形式。

例 （源代码：class_scope_refactor_1.coffee）

```
class Employee

  constructor: (@name, @dob)->

  printInfo: ->
    console.log "Name: #{@name}"
    console.log "DOB: #{@dob}"

emp1 = new Employee("Mark", new Date(1976, 7, 24))
emp1.printInfo()

emp2 = new Employee("Rachel", new Date(1979, 3, 28))
emp2.printInfo()
```

例 （源代码：class_scope_refactor_1.js）

```
(function() {
  var Employee, emp1, emp2;

  Employee = (function() {
    function Employee(name, dob) {
      this.name = name;
      this.dob = dob;
    }

    Employee.prototype.printInfo = function() {
```

```
    console.log("Name: " + this.name);
    return console.log("DOB: " + this.dob);
  };

  return Employee;

})();

emp1 = new Employee("Mark", new Date(1976, 7, 24));

emp1.printInfo();

emp2 = new Employee("Rachel", new Date(1979, 3, 28));

emp2.printInfo();

}).call(this);
```

输出 （源代码：class_scope_refactor_1.coffee）

```
Name: Mark
DOB: Tue Aug 24 1976 00:00:00 GMT-0400 (EDT)
Name: Rachel
DOB: Sat Apr 28 1979 00:00:00 GMT-0500 (EST)
```

现在代码更整洁，DRY[1]程度也更高了。当看到一个函数有两个参数时，我不禁要问，会不会有更多参数呢？如果有，该如何定义函数呢？在上述例子中，我担心的就是构造函数。如果要传入更多其他的参数，例如薪水、部门、经理，等等，该怎么办呢？我们来做些重构。

例 （源代码：class_scope_refactor_2.coffee）

```
class Employee

  constructor: (@attributes)->

  printInfo: ->
    console.log "Name: #{@attributes.name}"
    console.log "DOB: #{@attributes.dob}"

    if @attributes.salary
      console.log "Salary: #{@attributes.salary}"
    else
      console.log "Salary: Unknown"

emp1 = new Employee
  name: "Mark"
  dob: new Date(1976, 7, 24)
```

① http://en.wikipedia.org/wiki/DRY

```
  salary: "$1.00"

emp1.printInfo()

emp2 = new Employee
  name: "Rachel"
  dob: new Date(1979, 3, 28)

emp2.printInfo()
```

例（源代码: class_scope_refactor_2.js）

```
(function() {
  var Employee, emp1, emp2;

  Employee = (function() {

    function Employee(attributes) {
      this.attributes = attributes;
    }

    Employee.prototype.printInfo = function() {
      console.log("Name: " + this.attributes.name);
      console.log("DOB: " + this.attributes.dob);
      if (this.attributes.salary) {
        return console.log("Salary: " + this.attributes.salary);
      } else {
        return console.log("Salary: Unknown");
      }
    };

    return Employee;

  })();

  emp1 = new Employee({
    name: "Mark",
    dob: new Date(1976, 7, 24),
    salary: "$1.00"
  });

  emp1.printInfo();
  emp2 = new Employee({
    name: "Rachel",
    dob: new Date(1979, 3, 28)
  });

  emp2.printInfo();

}).call(this);
```

输出 （源代码：class_scope_refactor_2.coffee）

```
Name: Mark
DOB: Tue Aug 24 1976 00:00:00 GMT-0400 (EDT)
Salary: $1.00
Name: Rachel
DOB: Sat Apr 28 1979 00:00:00 GMT-0500 (EST)
Salary: Unknown
```

现在我们可以向 Employee 中传入任意多的参数了，它们都可以通过对象上的 attributes 属性访问到。这样更好而且扩展性更强。

如前例所示，我们给第一个员工传递了第三个属性——salary，而没有给第二个员工传入该参数。不过现在，我们传递上百个参数都没问题，代码不需要做任何修改。

关于这部分内容还有最后一点要介绍。有时，你可能会故作聪明。你可能会运用第 5 章中的知识，循环 attributes 属性的内容，将每一个键/值对直接赋值给对象，这样访问它们时，就不用再去访问@attributes 了，对吧？

好，我们来看看为什么说这是个非常糟糕的主意。

例 （源代码：class_scope_refactor_3.coffee）

```
class Employee

  constructor: (@attributes)->
    for key, value of @attributes
      @[key] = value

  printInfo: ->
    console.log "Name: #{@name}"
    console.log "DOB: #{@dob}"
    if @salary
      console.log "Salary: #{@salary}"
    else
      console.log "Salary: Unknown"

emp1 = new Employee
  name: "Mark"
  dob: new Date(1976, 7, 24)
  salary: "$1.00"

emp1.printInfo()

emp2 = new Employee
  name: "Rachel",
  dob: new Date(1979, 3, 28)
  printInfo: ->
    console.log "I've hacked your code!"

emp2.printInfo()
```

例（源代码: class_scope_refactor_3.js）

```
(function() {
  var Employee, emp1, emp2;

  Employee = (function() {

    function Employee(attributes) {
      var key, value, _ref;
      this.attributes = attributes;
      _ref = this.attributes;
      for (key in _ref) {
        value = _ref[key];
        this[key] = value;
      }
    }

    Employee.prototype.printInfo = function() {
      console.log("Name: " + this.name);
      console.log("DOB: " + this.dob);
      if (this.salary) {
        return console.log("Salary: " + this.salary);
      } else {
        return console.log("Salary: Unknown");
      }

    };

    return Employee;

  })();

  emp1 = new Employee({
    name: "Mark",
    dob: new Date(1976, 7, 24),
    salary: "$1.00"
  });

  emp1.printInfo();

  emp2 = new Employee({
    name: "Rachel",
    dob: new Date(1979, 3, 28),
    printInfo: function() {
      return console.log("I've hacked your code!");
    }
  });

  emp2.printInfo();

}).call(this);
```

输出 （源代码：class_scope_refactor_3.coffee）

```
Name: Mark
DOB: Tue Aug 24 1976 00:00:00 GMT-0400 (EDT)
Salary: $1.00
I've hacked your code!
```

在给第二个员工传递参数时，很容易就能把 printInfo 函数覆盖掉。这样的代码可读性很好，不过也很容易被攻击，谁愿意把代码写成这样呢？JavaScript 本身就很容易被篡改，我们为什么还要明知故犯呢？好了，了解了这点之后，我们继续介绍后续内容，就当我们没有做过这次重构。

6.5 扩展类

面向对象编程时，开发者经常需要用到继承[①]。继承允许在某个类（比如我们的 Employee 类）的基础上创建出其他的“变体”。

在实际场景中，每个人都是员工，但并非都是经理。因此，我们来定义一个 Manager 类，该类继承或者说扩展自 Employee 类。

例 （源代码：manager1.coffee）

```
class Employee

  constructor: (@attributes)->

  printInfo: ->
    console.log "Name: #{@attributes.name}"
    console.log "DOB: #{@attributes.dob}"
    console.log "Salary: #{@attributes.salary}"

class Manager extends Employee

employee = new Employee
  name: "Mark"
  dob: new Date(1976, 7, 24)
  salary: 50000

employee.printInfo()

manager = new Manager
  name: "Rachel"
  dob:new Date(1979, 3, 28)
  salary: 100000

manager.printInfo()
```

① http://en.wikipedia.org/wiki/Inheritance_(computer_science)

例 （源代码: manager1.js）

```
(function() {
  var Employee, Manager, employee, manager,
    __hasProp = Object.prototype.hasOwnProperty,
    __extends = function(child, parent) { for (var key in parent) { if
(__hasProp.call(parent, key)) child[key] = parent[key]; } function ctor() {
this.constructor = child; } ctor.prototype = parent.prototype; child.prototype =
new ctor; child.__super__ = parent.prototype; return child; };

  Employee = (function() {
    function Employee(attributes) {
      this.attributes = attributes;
    }

    Employee.prototype.printInfo = function() {
      console.log("Name: " + this.attributes.name);

      console.log("DOB: " + this.attributes.dob);
      return console.log("Salary: " + this.attributes.salary);
    };

    return Employee;

  })();

  Manager = (function(_super) {

    __extends(Manager, _super);

    function Manager() {
      Manager.__super__.constructor.apply(this, arguments);
    }

    return Manager;

  })(Employee);

  employee = new Employee({
    name: "Mark",
    dob: new Date(1976, 7, 24),
    salary: 50000
  });

  employee.printInfo();

  manager = new Manager({
    name: "Rachel",
    dob: new Date(1979, 3, 28),
    salary: 100000
```

```
  });

  manager.printInfo();

}).call(this);
```

输出 （源代码：manager1.coffee）

```
Name: Mark
DOB: Tue Aug 24 1976 00:00:00 GMT-0400 (EDT)
Salary: 50000
Name: Rachel
DOB: Sat Apr 28 1979 00:00:00 GMT-0500 (EST)
Salary: 100000
```

> **提示**：实际场景中，我们可能会用不同的角色来定义不同类型的员工，不过，这里为了讨论方便，假设这就是最好的解决多类员工的方式。

定义基本的 Manager 类很简单，class Manager 即可，不过，通过使用 extends 关键字，其后跟上要扩展的类名 Employee，我们就能够在 Manager 类中获得所有 Employee 类中的功能。

我们进一步来看看如何在子类中重写父类中的方法。我们给 Employee 类添加一个 bonus 函数，并返回 0。普通员工明显是没有奖金的，而经理有 10%的奖金，因此，要确保当调用经理的 bonus 函数时，能够返回正确的值。

例 （源代码：manager2.coffee）

```
class Employee

  constructor: (@attributes)->

  printInfo: ->
    console.log "Name: #{@attributes.name}"
    console.log "DOB: #{@attributes.dob}"
    console.log "Salary: #{@attributes.salary}"
    console.log "Bonus: #{@bonus()}"

  bonus: ->
    0

class Manager extends Employee

  bonus: ->
    @attributes.salary * .10

employee = new Employee
  name: "Mark"
```

```
  dob: new Date(1976, 7, 24)
  salary: 50000

employee.printInfo()

manager = new Manager
  name: "Rachel"
  dob: new Date(1979, 3, 28)
  salary: 100000

manager.printInfo()
```

例 （源代码: manager2.js）

```
(function() {
  var Employee, Manager, employee, manager,
    __hasProp = Object.prototype.hasOwnProperty,
    __extends = function(child, parent) { for (var key in parent) { if
(__hasProp.call(parent, key)) child[key] = parent[key]; } function ctor() {
this.constructor = child; } ctor.prototype = parent.prototype; child.prototype =
new ctor; child.__super__ = parent.prototype; return child; };

  Employee = (function() {

    function Employee(attributes) {
      this.attributes = attributes;
    }

    Employee.prototype.printInfo = function() {
      console.log("Name: " + this.attributes.name);
      console.log("DOB: " + this.attributes.dob);
      console.log("Salary: " + this.attributes.salary);
      return console.log("Bonus: " + (this.bonus()));
    };

    Employee.prototype.bonus = function() {
      return 0;
    };

    return Employee;

  })();

  Manager = (function(_super) {
    __extends(Manager, _super);

    function Manager() {
      Manager.__super__.constructor.apply(this, arguments);
    }

    Manager.prototype.bonus = function() {
```

```
      return this.attributes.salary * .10;
    };

    return Manager;

  })(Employee);

  employee = new Employee({
    name: "Mark",
    dob: new Date(1976, 7, 24),
    salary: 50000
  });

  employee.printInfo();

  manager = new Manager({
    name: "Rachel",
    dob: new Date(1979, 3, 28),
    salary: 100000
  });

  manager.printInfo();

}).call(this);
```

输出 （源代码：manager2.coffee）

```
Name: Mark
DOB: Tue Aug 24 1976 00:00:00 GMT-0400 (EDT)
Salary: 50000
Bonus: 0
Name: Rachel
DOB: Sat Apr 28 1979 00:00:00 GMT-0500 (EST)
Salary: 100000
Bonus: 10000
```

在子类中重载函数和重新定义一个函数一样，很简单。不过，如果想要调用原函数，并只添加一点额外的功能的话，要怎么做呢？请看下面这个例子。

在 printInfo 中看到奖金为 0，员工们一定很伤心，因此我们把打印奖金的信息从 printInfo 中移除掉，不过，还是要让经理看到奖金的。这个时候，我们就可以使用 super 关键字来处理。

例 （源代码：manager3.coffee）

```
class Employee

  constructor: (@attributes)->

  printInfo: ->
```

```
    console.log "Name: #{@attributes.name}"
    console.log "DOB: #{@attributes.dob}"
    console.log "Salary: #{@attributes.salary}"

  bonus: ->
    0

class Manager extends Employee

  bonus: ->
    @attributes.salary * .10

  printInfo: ->
    super
    console.log "Bonus: #{@bonus()}"

employee = new Employee
  name: "Mark"
  dob: new Date(1976, 7, 24)
  salary: 50000

employee.printInfo()

manager = new Manager
  name: "Rachel"
  dob: new Date(1979, 3, 28)
  salary: 100000

manager.printInfo()
```

例 （源代码：manager3.js）

```
(function() {
  var Employee, Manager, employee, manager,
    __hasProp = Object.prototype.hasOwnProperty,
    __extends = function(child, parent) { for (var key in parent) { if
(__hasProp.call(parent, key)) child[key] = parent[key]; } function ctor() {
this.constructor = child; } ctor.prototype = parent.prototype; child.prototype =
new ctor; child.__super__ = parent.prototype; return child; };

  Employee = (function() {

    function Employee(attributes) {
      this.attributes = attributes;
    }

    Employee.prototype.printInfo = function() {
      console.log("Name: " + this.attributes.name);
      console.log("DOB: " + this.attributes.dob);
      return console.log("Salary: " + this.attributes.salary);
    };
```

```
    Employee.prototype.bonus = function() {
      return 0;
    };

    return Employee;

  })();

  Manager = (function(_super) {

    __extends(Manager, _super);

    function Manager() {
      Manager.__super__.constructor.apply(this, arguments);
    }

    Manager.prototype.bonus = function() {
      return this.attributes.salary * .10;
    };

    Manager.prototype.printInfo = function() {
      Manager.__super__.printInfo.apply(this, arguments);
      return console.log("Bonus: " + (this.bonus()));
    };

    return Manager;

  })(Employee);

  employee = new Employee({
    name: "Mark",
    dob: new Date(1976, 7, 24),
    salary: 50000
  });

  employee.printInfo();

  manager = new Manager({
    name: "Rachel",
    dob: new Date(1979, 3, 28),
    salary: 100000
  });

  manager.printInfo();

}).call(this);
```

输出 （源代码：manager3.coffee）

```
Name: Mark
DOB: Tue Aug 24 1976 00:00:00 GMT-0400 (EDT)
```

```
Salary: 50000
Name: Rachel
DOB: Sat Apr 28 1979 00:00:00 GMT-0500 (EST)
Salary: 100000
Bonus: 10000
```

在 Manager 类中定义的 printInfo 函数中，首先调用 super。在调用 super 时，会调用 Employee 类中的 printInfo 函数，如果有参数，参数也可以传递进去。之后，我们调用 super 将经理的奖金信息打印出来。

提示：在任何重写的函数中都可以调用 super；并且不用非要在第一行调用（有些语言强制要求必须要在第一行调用 super）。甚至，可以在不需要的时候不调用 super。

提示：在调用 super 时，无需显式传递参数。在默认情况下，传递给重写函数中的参数会自动传递给 super 函数。不过，你要想显式地传递参数当然也没问题。有的时候，想要在调用 super 前将参数做一些调整，就需要显式地去传递修改后的参数。

6.6 类级函数

类级函数是不需要实例化就可以调用的函数。类级函数非常有用。最大的用处之一就是为函数提供了一种命名空间。比方说，JavaScript 中的 Math.random()。无须实例化 Math 对象就可以调用该方法获取随机数。将 random 函数挂在 Math 类上，可以有效避免因为其他地方也定义了 random 函数而导致将其覆盖的风险。

提示：事实上，Math 并非类，它只是一个用于定义命名空间的普通对象。这里之所以没有说得这么准确，也是为了更好地阐述我的观点。

你也可以用类级函数处理一些对该类实例产生影响的事情，比方说，在数据库中查询这些实例。

我们可以在 Employee 类上定义一些类级函数。因为没有数据库，所以我们就在创建对象时记录下对象数目，然后将其总数打印出来。为了记录雇佣的员工，我们创建一个叫 hire 的类级函数，用于将新雇佣的雇员添加到一个数组中，该数组用于扮演临时数据库的角色。与此同时，我们再创建一个 total 的类级函数，此函数返回所有“数据库”中的员工总数。

例（源代码：class_level.coffee）

```
class Employee

  constructor: ->
    Employee.hire(@)

  @hire: (employee) ->
```

```
    @allEmployees ||= []
    @allEmployees.push employee

  @total: ->
    console.log "There are #{@allEmployees.length} employees."
    @allEmployees.length

new Employee()
new Employee()
new Employee()

Employee.total()
```

例（源代码: class_level.js）

```
(function() {
  var Employee;

  Employee = (function() {

    function Employee() {
      Employee.hire(this);
    }

    Employee.hire = function(employee) {
      this.allEmployees || (this.allEmployees = []);
      return this.allEmployees.push(employee);
    };

    Employee.total = function() {
      console.log("There are " + this.allEmployees.length + " employees.");

      return this.allEmployees.length;
    };

    return Employee;

  })();

  new Employee();

  new Employee();

  new Employee();

  Employee.total();

}).call(this);
```

输出（源代码: class_level.coffee）

```
There are 3 employees.
```

那么我们是如何创建上述类级函数的呢？通过在函数名前使用@就等于告诉 CoffeeScript，此函数为类级函数。在 JavaScript 中也能照常工作，是因为@会转化为 this。这里 this 的上下文是 Employee 类，而不是该类的实例。

内部的类级函数作用域受限于其他的类级函数和属性。

> **提示**：我经常会定义一些只有类级方法的类。它们很适合用于构建工具包，可以确保函数有良好的作用域。并且，必要时还可以继承这些类，覆盖其中一些函数用于特殊需求。

不过，在定义类级函数和属性时，super 的使用是受限的。只有在子类函数中，想要调用被重写的函数时才可以使用 super。然而下面这个情况下用 super 的话问题就大了：如果调用函数时用的是 super，那么在访问类级属性时，就会严重出错。

我们来看看，如果我们在 Manager 类中重写 total 类级函数，并在函数内调用 super，会发生什么。

例（源代码：class_level_super.coffee）

```
class Employee

  constructor: ->
    Employee.hire(@)

  @hire: (employee) ->
    @allEmployees ||= []
    @allEmployees.push employee

  @total: ->
    console.log "There are #{@allEmployees.length} employees."
    @allEmployees.length

class Manager extends Employee

  @total: ->
    console.log "There are 0 managers."
    super

new Employee()
new Employee()
new Employee()

Manager.total()
```

例（源代码：class_level_super.js）

```
(function() {
  var Employee, Manager,
    __hasProp = Object.prototype.hasOwnProperty,
```

```
    __extends = function(child, parent) { for (var key in parent) { if
(_hasProp.call (parent, key)) child[key] = parent[key]; } function ctor() {
this.constructor = child; } ctor.prototype = parent.prototype; child.prototype =
new ctor; child.__super__ = parent.prototype; return child; };

  Employee = (function() {

    function Employee() {
      Employee.hire(this);
    }

    Employee.hire = function(employee) {
      this.allEmployees || (this.allEmployees = []);
      return this.allEmployees.push(employee);
    };

    Employee.total = function() {
      console.log("There are " + this.allEmployees.length + " employees.");
      return this.allEmployees.length;
    };

    return Employee;

  })();

  Manager = (function(_super) {

    __extends(Manager, _super);

    function Manager() {
      Manager.__super__.constructor.apply(this, arguments);
    }

    Manager.total = function() {
      console.log("There are 0 managers.");
      return Manager.__super__.constructor.total.apply(this, arguments);
    };

    return Manager;

  })(Employee);

  new Employee();

  new Employee();

  new Employee();

  Manager.total();

}).call(this);
```

输出 （源代码：class_level_super.coffee）

```
There are 0 managers.
TypeError: Cannot read property 'length' of undefined
    at Function.<anonymous> (.../classes/class_level_super.coffee:18:51)
    at Function.total (.../classes/class_level_super.coffee:36:50)
    at Object.<anonymous> (.../classes/class_level_super.coffee:49:11)
    at Object.<anonymous> (.../classes/class_level_super.coffee:51:4)
    at Module._compile (module.js:432:26)
    at Object.run (/usr/local/lib/node_modules/coffee-script/lib/coffee-script/
➥coffee-script.js:68:25)
    at /usr/local/lib/node_modules/coffee-script/lib/coffee-script/command.js:135:29
    at /usr/local/lib/node_modules/coffee-script/lib/coffee-script/command.js:110:18
    at [object Object].<anonymous> (fs.js:114:5)
    at [object Object].emit (events.js:64:17)
```

如上例子所示，当试图调用@allEmployees 属性上的 length 时，报了如下错误：

```
TypeError: Cannot read property 'length' of undefined
```

原因很简单，不过需要花点时间深入解释。此前介绍过，JavaScript 原生并不支持继承，CoffeeScript 模拟了类和继承。正因如此，子类 Manager 和 Employee 类完全是两个不同的对象，所以在 Employee 上定义的属性，在 Manager 类中根本不存在，Manager 类无法访问到。是不是有点儿晕？

> **提示：** 在类级函数中最好尽量避免使用 super。保持类级函数始终是自包含的，可以有效避免上述问题。

6.7 原型函数

在 JavaScript 中，要想在所有的对象实例上都添加函数或者属性，可以通过使用 prototype 属性将其添加到对象的原型上来实现。

在 CoffeeScript 中可以使用::操作符。我们来将 size 函数添加到所有的数组实例上，让 size 函数返回数组的长度。

例 （源代码：prototypes.coffee）

```
myArray = [1..10]

try
  console.log myArray.size()
catch error
  console.log error
```

```
Array::size = -> @length
console.log myArray.size()

myArray.push(11)
console.log myArray.size()
```

例 （源代码：prototypes.js）

```
(function() {
  var myArray;

  myArray = [1, 2, 3, 4, 5, 6, 7, 8, 9, 10];

  try {
    console.log(myArray.size());
  } catch (error) {
    console.log(error);
  }

  Array.prototype.size = function() {
    return this.length;
  };

  console.log(myArray.size());

  myArray.push(11);

  console.log(myArray.size());

}).call(this);
```

输出 （源代码：prototypes.coffee）

```
[TypeError: Object 1,2,3,4,5,6,7,8,9,10 has no method 'size']
10
11
```

因为数组默认并没有 size 函数，所以第一次调用 size 函数时报错了。但之后，我们在 Array 类的原型上添加了 size 函数，后面几次调用 size 函数就都没有问题了。

> **提示**：这里::操作符是种便捷操作。你仍然可以直接访问 prototype 属性，不过，有这么简洁的写法为什么还要写那么一长串呢？

6.8 绑定（->与=>）

JavaScript 是一门异步的[1]、事件驱动的编程语言。在非异步编程中，每次执行函数时，其

① http://en.wikipedia.org/wiki/Asynchronous_I/O

余代码都会被暂停，直到函数返回才会继续执行。在 JavaScript 中则无须如此。哪怕调用的函数还没有返回，依然可以继续执行后续代码。把这种编程方式想象为“一触即忘”或许有助于理解，程序一旦触发函数调用，就立马把它忘了，继续执行后续代码。我们来看一个异步的例子，看看程序是如何运行的。

例（源代码：fire_and_forget.coffee）

```
fire = (msg, wait)->
  setTimeout ->
    console.log msg
  , wait

fire("Hello", 3000)
fire("World", 1000)
```

例（源代码：fire_and_forget.js）

```
(function() {
  var fire;

  fire = function(msg, wait) {
    return setTimeout(function() {
      return console.log(msg);
    }, wait);
  };

  fire("Hello", 3000);

  fire("World", 1000);

}).call(this);
```

输出（源代码：fire_and_forget.coffee）

```
World
Hello
```

如上例所示，程序首先打印出“World”，然后才是“Hello”。在非异步情况下，结果肯定是“Hello”在前，几秒后再输出“World”。异步编程很强大，不过有点晦涩难懂。我们来看看什么情况下异步编程会变得很糟糕。

我们来写一个 log 方法。此方法会在控制台输出我们将要执行回调函数的日志信息；随后会执行回调函数；最后，会在控制台输出已经执行完回调函数的日志。

例（源代码：unbound.coffee）

```
class User

  constructor: (@name) ->
```

```
  sayHi: ->
    console.log "Hello #{@name}"

bob = new User('bob')
mary = new User('mary')

log = (callback)->
  console.log "about to execute callback..."
  callback()
  console.log "...executed callback"

log(bob.sayHi)
log(mary.sayHi)
```

例（源代码：unbound.js）

```
(function() {
  var User, bob, log, mary;

  User = (function() {

    function User(name) {
      this.name = name;
    }

    User.prototype.sayHi = function() {
      return console.log("Hello " + this.name);
    };

    return User;

  })();

  bob = new User('bob');

  mary = new User('mary');

  log = function(callback) {
    console.log("about to execute callback...");
    callback();
    return console.log("...executed callback");
  };

  log(bob.sayHi);

  log(mary.sayHi);

}).call(this);
```

输出 （源代码：unbound.coffee）

```
about to execute callback...
Hello undefined
...executed callback
about to execute callback...
Hello undefined
...executed callback
```

好吧，我确信这段代码不应该会有 undefined 的情况。那么，究竟发生了什么？在 log 函数中调用我们传递的回调函数时，回调函数已经丢失了原先的上下文，无法访问到我们在类中设置的 name 变量。这类问题在 JavaScript 中是很常见的，特别是在使用像 jQuery 这样的库发送 AJAX 请求或者绑定事件时。

那么如何解决这个问题呢？如何才能让回调函数回到原先的上下文呢？答案就是在 User 类中定义 sayHi 函数时使用=>（又叫粗箭头）来代替->。下面这个例子和此前的例子一样，唯一不同的就是将 sayHi：->改成了 sayHi：=>。我们来看看这次又会发生什么。

例 （源代码：bound.coffee）

```
class User

  constructor: (@name) ->

  sayHi: =>
    console.log "Hello #{@name}"

bob = new User('bob')
mary = new User('mary')

log = (callback)->
  console.log "about to execute callback..."
  callback()
  console.log "...executed callback"

log(bob.sayHi)
log(mary.sayHi)
```

例 （源代码：bound.js）

```
(function() {
  var User, bob, log, mary,
    __bind = function(fn, me){ return function(){ return fn.apply(me, arguments); }; };

  User = (function() {

    function User(name) {
      this.name = name;
      this.sayHi = __bind(this.sayHi, this);
```

```
    }

    User.prototype.sayHi = function() {
      return console.log("Hello " + this.name);
    };

    return User;

  })();

  bob = new User('bob');

  mary = new User('mary');

  log = function(callback) {
    console.log("about to execute callback...");
    callback();
    return console.log("...executed callback");
  };

  log(bob.sayHi);

  log(mary.sayHi);

}).call(this);
```

输出 （源代码：bound.coffee）

```
about to execute callback...
Hello bob
...executed callback
about to execute callback...
Hello mary
...executed callback
```

一个简单的字符带来的改变如此之大！为了更好地理解为什么一字之差区别如此之大，我们来对比下两个例子中编译后的unbound.js和bound.js代码。

例 （源代码：unbound.js）

```
(function() {
  var User, bob, log, mary;

  User = (function() {

    function User(name) {
      this.name = name;
    }

    User.prototype.sayHi = function() {
      return console.log("Hello " + this.name);
```

```
    };

    return User;

  })();

  bob = new User('bob');

  mary = new User('mary');

  log = function(callback) {
    console.log("about to execute callback...");
    callback();
    return console.log("...executed callback");
  };

  log(bob.sayHi);

  log(mary.sayHi);
}).call(this);
```

例（源代码: bound.js）

```
(function() {
  var User, bob, log, mary,
    __bind = function(fn, me){ return function(){ return fn.apply(me, arguments); }; };

  User = (function() {

    function User(name) {
      this.name = name;
      this.sayHi = __bind(this.sayHi, this);
    }

    User.prototype.sayHi = function() {
      return console.log("Hello " + this.name);
    };

    return User;

  })();

  bob = new User('bob');

  mary = new User('mary');

  log = function(callback) {
    console.log("about to execute callback...");
    callback();
    return console.log("...executed callback");
  };
```

```
    log(bob.sayHi);

    log(mary.sayHi);

  }).call(this);
```

在我解释上述两段代码区别前，如果你不明白 apply 函数的作用，那我建议你先找本 JavaScript 书把这部分内容先搞清楚。这两个 JavaScript 文件中有两处不同。第一处不同就是后者（使用了=>的例子）中有__bind 函数。该函数接收两个参数，第一个是要绑定的函数，第二个是要绑定该函数的上下文。__bind 函数会返回一个新函数，调用此新函数时，会用 apply 调用原先的函数，并将传递的上下文绑定到原先的函数上。

第二处不同是 User 类的构造函数不同。使用=>的代码中，会通过调用__bind 函数来重新定义 sayHi 函数，其中传入的参数是原先定义好的 sayHi 以及该类的实例。

如果你还搞不明白，也别担心——不明白的人不止你一个。上下文和上下文的绑定的确是 JavaScript 中比较难搞懂的内容。如果你搞不明白，我建议：第一，去读读对这块内容讲解得比较好的 JavaScript 书籍；第二，去读读 Yehuda 的 *Understanding JavaScript Function Invocation and "this"* [1]这篇博文，其中他对这方面的内容做了详细的解释，写得非常好。

如果你已经搞懂了，那现在你应该会很开心，因为，你知道了通过使用=>就再也不用去关心绑定的问题了。在本书第 11 章中还会在具体实践中使用=>。

6.9 小结

有意思吧？CoffeeScript 中的类是该语言最大的卖点之一，至少我是这样觉得的。我希望本章内容对你有所帮助。

本章介绍了很多内容。我们介绍了什么是 CoffeeScript 中的类，以及如何定义最基础的类。

随后，介绍了“特殊”的构造函数以及很多类中的作用域。

介绍了在 CoffeeScript 中如何使用 extends 关键字来扩展类。还介绍了如何使用 super 来帮助在子类中对父类函数功能做扩展。

在这之后，我们深入讲解了类级函数以及原型函数。还学习了在类级函数误 super 会导致的问题。

本章最后，介绍了相对比较复杂，但功能强大的概念——使用=>来绑定函数。

至此，你已经学了很多关于 CoffeeScript 的内容。恭喜！你可以学成“下山”了。本章是本书第一部分的最后一章。第二部分会将第一部分的知识运用到实践中，其中还会使用一些流行的 JavaScript 库。

① http://yehudakatz.com/2011/08/11/understanding-javascript-function-invocationand-this/

第二部分　CoffeeScript 实践

在本书第二部分中，会将第一部分介绍的 CoffeeScript 的理论知识运用到实践中。这样不仅有助于巩固对第一部分内容的理解，同时还希望在这过程中能学到些技巧。实践过程中，我们会采用一些流行的在 CoffeeScript 中常用的工具。

我们会用 Cake 文件（Cakefile）来运行应用中常用的任务。还会介绍 Jasmine 这套测试框架来帮助更容易地测试 CoffeeScript 代码。之后，会简单介绍 Node.js 并用它来构建一个应用服务器用于编译 CoffeeScript 代码。

最后，我们会构建一个完整的应用——经典的“待办事宜”（todo）应用。我们会用 3 章的内容来介绍如何构建该应用。首先，构建服务器端部分，负责待办事宜的持久化。其次，使用流行的 jQuery 库来构建客户端部分。最后，采用 Backbone[①]框架来替换掉手写的 jQuery[②]代码。该应用会采用一些比较有意思的库：jQuery、Backbone.js、Mongoose[③]、Express[④]，当然还有 CoffeeScript。

在使用这些不同技术来构建我们的应用和示例的过程中，我们不会深入去介绍这些库。只会强调用到的部分。所以，总的来说，本书这部分内容非常有趣！还等什么，让我们开始吧！

---

① http://documentcloud.github.com/backbone/
② http://jquery.com/
③ http://mongoosejs.com/
④ http://expressjs.com/

第 7 章

Cake 与 Cakefile

CoffeeScript 提供了一个叫 Cake 的简单构建工具，其本质和 Ruby 的 Rake[1]工具很类似。Cake 允许定义一些简单的任务来帮助构建 CoffeeScript 项目。比方说，运行测试的任务或者打包文件的任务。有了 Cake，定义这些任务就简单了，只需在一个叫 Cakefile 的文件中定义就行了。

本章将介绍如何定义和执行 Cake 任务，当然是使用 CoffeeScript 来做这些事。

> **提示：**"等等，我们要如何安装 Cake 呢？"根本不需要安装。安装了 CoffeeScript 就自动安装了 Cake 和它的命令行工具，也叫 `cake`。无须任何额外工作就能使用 Cake 了。

7.1 从这里开始

动手写第一个 Cake 任务前，我们先要弄明白和 Cake 如何工作相关的几件事情，这很重要。首先，所有的任务必须要定义在名为 Cakefile 的文件中，并且 Cakefile 文件必须在要运行 Cake 任务的目录中，通常是在项目的根目录中。

其次，需要知道的就是 Cakefile 必须得用 CoffeeScript 书写。Cakefile 内置了一些特殊的函数来帮助书写任务，具体会在我们创建 Cake 任务时做介绍。

7.2 创建 Cake 任务

我们来构建第一个 Cake 任务——一个简单的"hello world"任务。

① https://github.com/jimweirich/rake

例（源代码：example1/Cakefile）

```
task "greet", "Say hi to the nice people", ->
  console.log "Hello, World!"
```

例（源代码：example1/Cakefile.js）

```
(function() {

  task("greet", "Say hi to the nice people", function() {
    return console.log("Hello, World!");
  });

}).call(this);
```

> **提示**：尽管 Cakefile 是在后台编译 JavaScript 代码的，你根本看不到编译后的 JavaScript 代码，不过我还是决定将编译后的 JavaScript 代码展示出来，以帮助我们更好地理解 CoffeeScript 背后的原理。

要定义 Cake 任务，需要调用 task 函数，该函数在每个 Cakefile 中都默认支持。task 函数的第一个参数是任务名，该任务名会在命令行执行任务时使用。第二个参数是对任务的描述，是个可选参数。如果添加了该参数，那么在打印出任务列表时，任务描述信息也会打印出来。task 函数的最后一个参数是一个函数，在执行任务时会被调用。函数内容其实就是任务本身。

使用 cake 命令行工具可以查看任务列表：

```
>cake
```

在 Cakefile 所在目录运行 cake 命令，会看到如下输出。

输出（源代码：example1/Cakefile）

```
Cakefile defines the following tasks:

cake greet                # Say hi to the nice people
```

如上述结果所示，cake 命令后紧跟着我们定义的任务的名字，随后是任务的描述。

> **提示**：当你使用一些库，并且这些库自带一些 Cake 任务的话，输出所有的任务列表是非常有用的。

7.3　执行 Cake 任务

我们已经创建了第一个任务，那么该如何执行任务呢？其实在运行例子的 cake 命令时，就已经提示该如何运行了。简单地输入 cake 命令，然后加上要执行的任务名即可，就像下面这样：

```
> cake greet
```

这样就会运行我们定义的 greet 任务了。

输出

```
Hello, World!
```

7.4　使用选项

我们已经书写了第一个任务，并且也知道了如何执行它，但如何给任务传递参数呢？比方说，我们想要给 greet 任务传递一个选项，以便可以自定义要打招呼的人，该怎么做呢？让我们来看看。

要给任务传递选项，首先第一步就是定义选项。通过使用 Cake 提供的 option 函数来定义选项。该函数接收 3 个参数。第一个参数代表选项的“简写”形式，第二个参数代表选项的“完整”形式，最后一个参数代表该选项的简单描述。让我们再来看看 greet 任务，这次我们来定义一个选项以便我们能自定义打招呼。

> **提示：**这里所说的选项“简写”形式和“完整”形式，指的是用户要敲的选项内容长短。比方说，-n 就是简写形式，而--name 就是完整形式。

例（源代码：example2/Cakefile）

```
option '-n', '--name [NAME]', 'name you want to greet'
task "greet", "Say hi to someone", (options)->
  message = "Hello, "
  if options.name?
    message += options.name
  else
    message += "World"
  console.log message
```

例（源代码：example2/Cakefile.js）

```
(function() {

  option('-n', '--name [NAME]', 'name you want to greet');

  task("greet", "Say hi to someone", function(options) {
    var message;
    message = "Hello, ";
```

```
    if (options.name != null) {
      message += options.name;
    } else {
      message += "World";
    }
    return console.log(message);
  });

}).call(this);
```

如上述代码所示，在定义任务之前，我们调用了 option 函数，并将 3 个预期的参数传递给它。第一个参数是-n，是选项的简写形式。第二个参数是--name，是选项的完整形式。在这个例子中要注意的是，选项完整形式中，还有[NAME]，这等于告诉 Cake，这里需要一个值。如果这里没有写明像[NAME]这样的形式，那么，当你试图给选项传递值的时候 Cake 就会报错。最后一个参数是对选项的描述。

> **提示：** 尽管这 3 个参数都是 option 函数要求的，但是事实上只有最后两个参数才是必要的。选项的完整形式和描述都是必须的，而选项的简写形式可以不指定。如果你不需要选项的简写形式，那第一个参数传递一个空的字符串或者 null 就可以了。

现在当我们再次运行 cake 命令查看可用任务列表时，会得到如下输出。

输出 （源代码：example2/Cakefile）

```
Cakefile defines the following tasks:

cake greet          # Say hi to someone

  -n, --name   name you want to greet
```

输出内容的最后可以看到任务相应的可用选项。

> **提示：** 这里我要指出我个人认为的 Cake 选项的缺点。Cake 选项不是针对特定的任务的，而是针对所有的任务的。也就是说，除了 greet 任务之外，如果我们还有另外一个任务，那么这两个任务都可以接收 name 选项。尽管这不是什么特别严重的问题，不过这就意味着在定义选项名字和描述的时候要多留心。

再仔细看一下我们的 greet 任务，你会发现我们传递给任务函数的参数是一个对象(options)。

例 （源代码：example2/Cakefile）

```
option '-n', '--name [NAME]', 'name you want to greet'
task "greet", "Say hi to someone", (options)->
  message = "Hello, "
  if options.name?
```

```
    message += options.name
  else
    message += "World"
  console.log message
```

例　（源代码：example2/Cakefile.js）

```
(function() {

  option('-n', '--name [NAME]', 'name you want to greet');

  task("greet", "Say hi to someone", function(options) {
    var message;
    message = "Hello, ";
    if (options.name != null) {
      message += options.name;
    } else {
      message += "World";
    }
    return console.log(message);
  });

}).call(this);
```

通过 options 对象，我们就能判断执行任务时有没有传递 name 选项。如果传了 name 选项，就跟指定名字的人打招呼；否则，直接输出“Hello World”。

我们可以在调用执行 greet 任务时指定 name 选项，如下所示。

```
> cake -n Mark greet
```

输出　（源代码：example2a/Cakefile）

```
Hello, Mark
```

要想让选项成为必选参数，得要自己手动去检查任务定义看该选项是否存在；如果不存在则抛出错误。

例　（源代码：example2a/Cakefile）

```
option '-n', '--name [NAME]', 'name you want to greet'
task "greet", "Say hi to someone", (options)->
  throw new Error("[NAME] is required") unless options.name?
  console.log "Hello, #{options.name}"
```

例　（源代码：example2a/Cakefile.js）

```
(function() {

  option('-n', '--name [NAME]', 'name you want to greet');
```

```
  task("greet", "Say hi to someone", function(options) {
    if (options.name == null) throw new Error("[NAME] is required");
    return console.log("Hello, " + options.name);
  });

}).call(this);
```

输出（源代码：example2a/Cakefile）

```
node.js:201
        throw e; // process.nextTick error, or 'error' event on first tick
              ^
Error: [NAME] is required
    at Object.action (.../cake/example2a/Cakefile:6:37)
    at /usr/local/lib/node_modules/coffee-script/lib/coffee-script/cake.js:39:26
    at Object.run (/usr/local/lib/node_modules/coffee-script/lib/coffee-script/
➥cake.js:62:21)
    at Object.<anonymous> (/usr/local/lib/node_modules/coffee-script/bin/cake:7:38)
    at Module._compile (module.js:432:26)
    at Object..js (module.js:450:10)
    at Module.load (module.js:351:31)
    at Function._load (module.js:310:12)
    at Array.0 (module.js:470:10)
    at EventEmitter._tickCallback (node.js:192:40)
```

> **提示**：在运行带选项的 Cake 任务时，要谨记所有的选项都必须放在任务名之前。否则，就会得到非常不友好的错误。对我而言，这种方式并不好。我个人更倾向于 cake greet -n Mark 这种方式，不过遗憾的是，截止到本书撰写时，还不支持这种方式。

7.5 调用其他任务

有时会有在任务中调用另外一个任务的需求。比方说：一个项目通常会有这样两个任务，一个任务是将构建目录清空，另外一个任务是编译和构建项目。我们来定义一下这两个任务。

例（源代码：example3/Cakefile）

```
task "clean", "Clean up build directories", ->
  console.log "cleaning up..."

task "build", "Build the project files", ->
  console.log "building..."
```

例 （源代码：example3/Cakefile.js）

```
(function() {

  task("clean", "Clean up build directories", function() {
    return console.log("cleaning up...");
  });

  task("build", "Build the project files", function() {
    return console.log("building...");
  });

}).call(this);
```

输出 （源代码：example3/Cakefile）

```
Cakefile defines the following tasks:

cake clean          # Clean up build directories
cake build          # Build the project files
```

使用这两个任务一段时间后，你会发现经常需要运行 clean 任务，然后再运行 build 任务。这个时候，你可以通过一条命令同时运行两个任务，如下所示。

```
> cake clean build
```

这条命令会执行两个任务。不过，如果这个时候还有第 3 个打包的任务 package，怎么办呢？在打包项目之前，你想要先确保已经构建了，在构建之前，你又想要确保构建目录是干净的。这个时候，你可以采用如下这种方式。

```
> cake clean build package
```

这种方式问题在于很容易出错。如果忘记了先执行 build 和 clean 任务，怎么办？会发生什么？幸运的是，Cake 允许在一个任务中调用另一个任务。通过使用 Cake 提供的 invoke 函数，并将要调用的任务名传递给该函数就能实现调用任务的目的了。

例 （源代码：example4/Cakefile）

```
task "clean", "Clean up build directories", ->
  console.log "cleaning up..."

task "build", "Build the project files", ->
  console.log "building..."

task "package", "Clean, build, and package the project", ->
  invoke "clean"
```

```
  invoke "build"
  console.log "packaging..."
```

例 （源代码：example4/Cakefile.js）

```
(function() {

  task("clean", "Clean up build directories", function() {
    return console.log("cleaning up...");
  });

  task("build", "Build the project files", function() {
    return console.log("building...");
  });

  task("package", "Clean, build, and package the project", function() {
    invoke("clean");
    invoke("build");
    return console.log("packaging...");
  });

}).call(this);
```

输出 （源代码：example4/Cakefile）

```
Cakefile defines the following tasks:

cake clean           # Clean up build directories
cake build           # Build the project files
cake package         # Clean, build, and package the project
```

现在我们可以直接调用 package 任务，相应的 clean 任务和 build 任务都会被执行。

```
> cake package
```

输出 （源代码：example4/Cakefile）

```
cleaning up...
building...
packaging...
```

> **提示**：在调用其他任务时，有很重要的一点要记住，这些任务都会被异步执行。因此，在我们的例子中，是无法保证 clean 任务一定会在 build 任务执行前完成。这会导致一些潜在的问题，要小心。同样地，此前介绍的利用一条 cake 命令执行多个任务也会有类似的问题。

7.6　小结

本章介绍了 CoffeeScript 内置的称为 Cake 的构建工具，介绍了如何定义任务以及查看任务，还介绍了如何执行任务、给任务添加选项以及同时执行多个任务。可能你会觉得没有真正学到很多关于如何编写有用的 Cake 任务的内容。原因在于 Cake 本身就是如此。写什么样的任务取决于你自己的需求。如果要找一些有用的模块，用来帮助完成读写文件以及目录、编译 CoffeeScript 文件、发送 HTTP 请求等的任务，那么第 9 章中要介绍的 Node.js[①]项目是个很好的选择。其他章节会介绍诸如运行测试之类的任务，要注意相应的例子。

Cake 的确是个很好的工具，并且随着 CoffeeScript 的安装自动就安装好了，非常方便，不过，我发现它对于我想要书写的大部分任务来说还是太弱了，而且用法很不方便。当结合 Node[②]的模块以及其他可用模块一同使用时，它可以更强大，不过，我发现自己仍然更倾向于使用 Rake，因为与 Cake 相比它更优雅。

不过，我强烈建议你用一下 Cake。或许，它很适合你，特别是当你能使用的语言以及安装的语言有限的情况下，它的确很不错。

---

① http://nodejs.org/
② http://nodejs.org/

第 8 章

使用 Jasmine 测试

对我来说，没有比写代码不测试更罪恶的事情了。我深信测试绝不是可选的[①]，而是必需的。特别是在我实践了 TDD[②]之后，感觉越发强烈。TDD，即测试驱动开发（test-driven development）,是一种编程的方法论，它提倡先写测试再写代码。通过实践 TDD 之后，有助于提高睡眠质量，因为你知道代码都被测试过了，并且有问题也能及早地解决，不用提心吊胆了。

TDD 有时会让人望而生畏；很多人不知道从哪里开始。如果你需要一些如何成为一名测试驱动的开发者的指导，请允许我向你推荐一篇名为《如何成为一名测试驱动开发者》（How to Become a Test-driven Developer）的博文[③]。

本章会为大家简单介绍我认为最好的测试 JavaScript 应用的测试工具之一——Jasmine[④]。

提示：事实上，Jasmine 并非 JavaScript 唯一的测试工具，还有其他一些类似的工具。我之所以认为它是最好的工具之一，是因为它借鉴了我最喜欢的 Ruby 测试框架——RSpec[⑤]。其他 JavaScript 的测试框架还有 Qunit[⑥]、JsTestDriver[⑦]及 YUI Test[⑧]。

这里不会对 Jasmine 做非常详细的介绍，网上可以找到一堆详细介绍 Jasmine 的文章和视频。这里我要介绍的是如何在 CoffeeScript 中使用 Jasmine，让你对 Jasmine 的功能有一个大概的了解。

① http://www.metabates.com/2010/07/01/testing-is-not-an-option/
② http://en.wikipedia.org/wiki/Test-driven_development
③ http://www.metabates.com/2010/10/12/how-to-become-a-test-driven-developer/
④ http://pivotal.github.com/jasmine/
⑤ https://github.com/rspec/rspec
⑥ http://docs.jquery.com/QUnit
⑦ http://code.google.com/p/js-test-driver/
⑧ http://yuilibrary.com/yui/docs/test/

8.1 安装 Jasmine

你可能还记得本书最开始提到过，我不太喜欢在书中介绍工具的安装。原因很明显，很有可能等到书出版后，其安装指令就变了。因此，这里也不会讲安装 Jasmine 的具体方式。

不用担心，我会介绍我认为最好的在 CoffeeScript 中安装和准备 Jasmine 环境的方式。这种方式要用到一个叫 jasmine-headless-webkit[①]的 Ruby gem。因为是个 Ruby gem，所以要求先安装 Ruby。除此之外，还有其他的一些依赖条件。脚注中的链接有详细的指南。

准备 Jasmine 环境的方式并不复杂，不过，都不支持原生 CoffeeScript，都要求提前将所有的代码源文件和测试文件编译好，说实在的，谁会愿意这样干呢？

如果你不想用 jasmine-headless-webkit，想使用另外版本的 Jasmine，也可以。不过本章介绍过程中会使用 jasmine-headless-webkit，所以如果你打算学习本章内容的话，我建议你还是使用 jasmine-headless-webkit。

8.2 准备 Jasmine 环境

假定你已经安装好 Jasmine 并一切就绪了，那就开始吧。

本章我们会构建一个简单的计算器项目。实现的计算器会做一些简单的运算：加、减、乘、除。首先，创建一个新的项目文件夹。到该文件夹目录中运行如下命令来准备 Jasmine 环境：

```
> jasmine init
```

执行完上述命令后，Jasmine 应该已经在项目目录中创建了一堆新的文件和文件夹，如下所示：

```
public/
  javascripts/
    Player.js
    Song.js
Rakefile
spec/
  javascripts/
    helpers/
      SpecHelper.js
    PlayerSpec.js
  support/
    jasmine.yml
    jasmine_config.rb
    jasmine_runner.rb
```

Jasmine 同时也创建了一些示例文件，一方面帮助你理解它是如何工作的，另外一方面也帮

① http://johnbintz.github.com/jasmine-headless-webkit/

助你确认 Jasmine 环境已经正确地搭建好了。接下来我们就来测试下：

```
> jasmine-headless-webkit -c
```

顺利的话，你应该会看到类似如下的输出内容：

```
Running Jasmine specs...
.....
PASS: 5 tests, 0 failures, 0.009 secs.
```

太好了！现在已经将 Jasmine 运行起来了。看起来不错，除了每次想要运行测试用例时，要在命令行执行的命令有点恶心之外，让我们用第 7 章学到的内容，写个简单的 Cake 任务来简化操作。

先来看下 Cakefile，后面再对此做解释。

例（源代码：calc.1/Cakefile）

```
exec = require('child_process').exec

task "test", (options) =>
  exec "jasmine-headless-webkit -c", (error, stdout, stderr)->
    console.log stdout
```

例（源代码：calc.1/Cakefile.js）

```
(function() {
  var exec,
    _this = this;

  exec = require('child_process').exec;

  task("test", function(options) {
    return exec("jasmine-headless-webkit -c", function(error, stdout, stderr) {
      return console.log(stdout);
    });
  });

}).call(this);
```

输出（源代码：calc.1/Cakefile）

```
Cakefile defines the following tasks:

cake test
```

这绝不算你要看到的最复杂的 Cakefile 或者 Cake 任务，不过，还是值得对其做一番解释。上述 Cake 任务最神奇的地方发生在第一行，在这行我们导入了 Node.js[①]中的 child_process 模块。如果你对 Node.js 以及该模块还不了解，不用担心，我们会在第 9 章中做相应介绍。

① http://nodejs.org/

我们主要在 Cake 任务中使用了 child_process 的 exec 函数。exec 函数允许往控制台发送命令，然后捕获命令结束后的输出结果。

我们创建了一个名为 test 的 Cake 任务来向控制台发送 jasmine-headless-webkit-c 的命令，来运行测试用例。运行结束后，会执行我们的回调函数，将运行结果打印到控制台。

现在，如果你在控制台输入：

```
> cake test
```

你会看到和此前一样的结果。

简化了命令行之后，在正式开始写测试用例之前，我们来为项目做些配置。

首先，我们将那些生成好的我们不需要的示例文件移除，让我们的项目更为整洁。最终，项目路径如下所示。

```
src/
Rakefile
spec/
  javascripts/
    helpers/
   support/
     jasmine.yml
     jasmine_config.rb
     jasmine_runner.rb
```

好，差不多了。最后要做的就是修改配置文件 spec/javascripts/support/jasmine.yml 为如下形式。

例 （源代码：calc.2/spec/javascripts/support/jasmine.yml）

```
src_files:
  - "**/*.coffee"

helpers:
     - "helpers/**/*.coffee"

spec_files:
  - /**/*_spec.coffee

src_dir: "src"

spec_dir: spec/javascripts
```

现在，我们终于可以开始介绍如何用 Jasmine 编写测试用例了。

8.3 Jasmine 介绍

用 Jasmine 写的测试用例是什么样的呢？我们来写一个简单的来一探究竟吧。

在 spec 目录中，我们来创建一个名为 calculator_spec.coffee 的新文件。该文件用来存放所有计算器的测试用例。先来看看一个简单的测试用例长什么样，然后再对其做解释。

例 （源代码: calc.3/spec/javascripts/calculator_spec.coffee）

```
describe "Calculator", ->

  it "does something", ->
    expect(1 + 1).toEqual 2
    expect(1 + 1).not.toEqual 3
```

例 （源代码: calc.3/spec/javascripts/calculator_spec.js）

```
(function() {

  describe("Calculator", function() {
    return it("does something", function() {
      expect(1 + 1).toEqual(2);
      return expect(1 + 1).not.toEqual(3);
    });
  });

}).call(this);
```

首先，我们需要创建一个名为"describe"的代码块，以此告诉 Jasmine 要测试的对象是"谁"。通常这里的"谁"会是类或者函数。上述例子中是 Calculator 类。describe 函数的第一个参数是一个字符串，上述例子中是"Calculator"。第二个参数是个函数，包含了与第一个参数指定的对象相关的所有测试用例。

在传递给 describe 函数的回调函数中，可以定义"it"代码块。"it"代码块其实就是一个函数，和"describe"函数类似。它也接收两个参数。第一个参数是字符串，表示对测试内容的描述。上述例子中，我们想要测试"Calculator""does something"。第二个参数是一个函数，包含了测试断言。

在我们的"does something"测试用例中，我们断言 1+1 等于 2。我们还得意地断言 1+1 不等于 3。我们用匹配器来做这些测试；上述例子中，toEqual 就是一台匹配器。这台匹配器只有一条规则：如果匹配成功返回 true，则测试通过；否则，测试失败。

那么 expect 函数用来干什么呢？expect 函数接收我们要测试的语句作为参数，针对上述例子就是 1+1，然后返回一个包含匹配器函数的特殊对象。简单来说，expect 简化了 Jasmine 用例代码的编写，同时也更具可读性。

> **提示**：Jasmine 内置了一些非常有用的匹配器，可以用它们来匹配任何场景。通过访问 Jasmine 官网的文档[①]可以查看所有这些匹配器的列表。

8.4 单元测试

我们已经对如何用 Jasmine 写测试用例有了基本的了解，下面来进一步完善对 Calculator

① https://github.com/pivotal/jasmine/wiki/Matchers

类的测试用例。移除此前的示例用例，增加 4 个 describe 代码块来测试加、减、乘、除。

例（源代码：calc.4/spec/javascripts/calculator_spec.coffee）

```
describe "Calculator", ->

  describe "#add", ->

    it "adds two numbers", ->

  describe "#subtract", ->

    it "subtracts two numbers", ->

  describe "#multiply", ->

    it "multiplies two numbers", ->

  describe "#divide", ->

    it "divides two numbers", ->
```

例（源代码：calc.4/spec/javascripts/calculator_spec.js）

```
(function() {

  describe("Calculator", function() {
    describe("#add", function() {
      return it("adds two numbers", function() {});
    });
    describe("#subtract", function() {
      return it("subtracts two numbers", function() {});
    });
    describe("#multiply", function() {
      return it("multiplies two numbers", function() {});
    });
    return describe("#divide", function() {
      return it("divides two numbers", function() {});
    });
  });

}).call(this);
```

你可能会对上述代码有这样的疑惑：为什么在“describe”代码块中还有“describe”代码块？那是因为，我们要对 Calculator 类中的 4 个函数分别写测试用例。每个函数对应了 describe 中的第一个参数，如"#add"、"#subtract"等，describe 中的“it”代码块则包含了我们真正要测试的内容。

我们可以通过如下命令来运行测试用例：

```
> cake test
```

其输出结果如下所示：

```
Running Jasmine specs...
....
PASS: 4 tests, 0 failures, 0.02 secs.
```

> **提示：**在我们的测试用例中，每个函数名前都加了#，这其实是种通用的测试代码的风格，表示该函数是实例函数。如果要表示类函数的话，函数名前要加的就是.；而不是#了。

之所以所有的测试用例都通过了是因为我们并没有在“it”代码块中添加断言。下面我们就来添加一些。

例（源代码：calc.5/spec/javascripts/calculator_spec.coffee）

```
describe "Calculator", ->

  describe "#add", ->

    it "adds two numbers", ->
      calculator = new Calculator()
      expect(calculator.add(1, 1)).toEqual 2

  describe "#subtract", ->

    it "subtracts two numbers", ->
      calculator = new Calculator()
      expect(calculator.subtract(10, 1)).toEqual 9

  describe "#multiply", ->

    it "multiplies two numbers", ->
      calculator = new Calculator()
      expect(calculator.multiply(5, 4)).toEqual 20

  describe "#divide", ->

    it "divides two numbers", ->
      calculator = new Calculator()
      expect(calculator.divide(20, 5)).toEqual 4
```

例（源代码：calc.5/spec/javascripts/calculator_spec.js）

```
(function() {

  describe("Calculator", function() {
```

```
describe("#add", function() {
  return it("adds two numbers", function() {
    var calculator;
    calculator = new Calculator();
    return expect(calculator.add(1, 1)).toEqual(2);
  });
});
describe("#subtract", function() {
  return it("subtracts two numbers", function() {
    var calculator;
    calculator = new Calculator();
    return expect(calculator.subtract(10, 1)).toEqual(9);
  });
});
describe("#multiply", function() {
  return it("multiplies two numbers", function() {
    var calculator;
    calculator = new Calculator();
    return expect(calculator.multiply(5, 4)).toEqual(20);
  });
});
return describe("#divide", function() {
  return it("divides two numbers", function() {
    var calculator;
    calculator = new Calculator();
    return expect(calculator.divide(20, 5)).toEqual(4);
  });
});
});

}).call(this);
```

现在，渐渐成形了。我们有了不错的测试用例，用来测试要在 Calculator 类中定义的 4 个函数，那么，运行这些测试用例后的结果是什么样的呢？

```
Running Jasmine specs...
FFFF

Calculator #add adds two numbers.
➥(../jasmine/calc.5/spec/javascripts/calculator_spec.coffee:5)
  ReferenceError: Can't find variable: Calculator in ../jasmine/calc.5/spec/
➥javascripts/calculator_spec.coffee (line ~6)

Calculator #subtract subtracts two numbers.
➥(../jasmine/calc.5/spec/javascripts/calculator_spec.coffee:11)
  ReferenceError: Can't find variable: Calculator in ../jasmine/calc.5/spec/
➥javascripts/calculator_spec.coffee (line ~13)
```

```
Calculator #multiply multiplies two numbers.
➥(../jasmine/calc.5/spec/javascripts/calculator_spec.coffee:17)
  ReferenceError: Can't find variable: Calculator in ../jasmine/calc.5/spec/
➥javascripts/calculator_spec.coffee (line ~20)

Calculator #divide divides two numbers.
➥(../jasmine/calc.5/spec/javascripts/calculator_spec.coffee:23)
  ReferenceError: Can't find variable: Calculator in ../jasmine/calc.5/spec/
➥javascripts/calculator_spec.coffee (line ~27)

FAIL: 4 tests, 4 failures, 0.017 secs.
```

所有的测试用例都失败了，失败就对了，因为我们压根儿就还没有实现 Calculator 类呢！那么下面就来定义 Calculator 类，这非常简单。

例（源代码：calc.6/src/calculator.coffee）

```
class @Calculator

  add: (a, b) ->
    a + b

  subtract: (a, b) ->
    a - b

  multiply: (a, b) ->
    a * b

  divide: (a, b) ->
    a / b
```

例（源代码：calc.6/src/calculator.js）

```
(function() {

  this.Calculator = (function() {

    function Calculator() {}

    Calculator.prototype.add = function(a, b) {
      return a + b;
    };

    Calculator.prototype.subtract = function(a, b) {
      return a - b;
    };

    Calculator.prototype.multiply = function(a, b) {
      return a * b;
    };
```

```
    Calculator.prototype.divide = function(a, b) {
      return a / b;
    };

    return Calculator;
  })();

}).call(this);
```

现在再来运行测试用例时，就会发现都通过了：

```
Running Jasmine specs...
....
PASS: 4 tests, 0 failures, 0.021 secs.
```

8.5 Before 与 After

尽管我们的测试用例看起来不错了，不过，每个用例中还有很多重复的代码。每个测试用例中，我们都创建了一个 Calculator 类的实例。这个时候，我们可以使用 Jasmine 提供的 beforeEach 函数来去除重复代码。

> **提示**：你可能在猜想，既然有 beforeEach 函数，那是不是对应还有 afterEach 函数呢？确实有。afterEach 函数对于重置数据库、重置文件或者恢复测试前的数据等都很有帮助。

我们来将创建 Calculator 类实例的操作移到 beforeEach 函数中。只要给 beforeEach 函数传递一个函数即可，该函数会在当前“describe”代码块中，甚至是子级“describe”代码块中的每个“it”代码块运行前被执行。

例（源代码：calc.7/spec/javascripts/calculator_spec.coffee）

```
describe "Calculator", ->

  beforeEach ->
    @calculator = new Calculator()

  describe "#add", ->

    it "adds two numbers", ->
      expect(@calculator.add(1, 1)).toEqual 2

  describe "#subtract", ->

    it "subtracts two numbers", ->
      expect(@calculator.subtract(10, 1)).toEqual 9
```

```
  describe "#multiply", ->

    it "multiplies two numbers", ->
      expect(@calculator.multiply(5, 4)).toEqual 20

  describe "#divide", ->

    it "divides two numbers", ->
      expect(@calculator.divide(20, 5)).toEqual 4
```

例（源代码：calc.7/spec/javascripts/calculator_spec.js）

```
(function() {

  describe("Calculator", function() {
    beforeEach(function() {
      return this.calculator = new Calculator();
    });
    describe("#add", function() {
      return it("adds two numbers", function() {
        return expect(this.calculator.add(1, 1)).toEqual(2);
      });
    });
    describe("#subtract", function() {
      return it("subtracts two numbers", function() {
        return expect(this.calculator.subtract(10, 1)).toEqual(9);
      });
    });
    describe("#multiply", function() {
      return it("multiplies two numbers", function() {
        return expect(this.calculator.multiply(5, 4)).toEqual(20);
      });
    });
    return describe("#divide", function() {
      return it("divides two numbers", function() {
        return expect(this.calculator.divide(20, 5)).toEqual(4);
      });
    });
  });

}).call(this);
```

> **提示**：对于 beforeEach 和 afterEach 的作用域可能会有点困惑，不太好理解。你可以把它想象成瀑布。从当前“describe”代码块的作用域一直“流”到所有子级，甚至多级嵌套的“describe”代码块中。

在编写 beforeEach 函数时，有一点很重要：你可以写任意多个 beforeEach 函数，并且可以写在任意地方。继续用此前的 Calculator 类来做实践。

我们给 Calculator 加一个标志，来表示计算器是否应该在科学计算模式下进行运算。

例 （源代码: calc.8/src/calculator.coffee）

```
class @Calculator

  constructor: (@scientific = false)->

  add: (a, b) ->
    a + b

  subtract: (a, b) ->
    a - b

  multiply: (a, b) ->
    a * b

  divide: (a, b) ->
    a / b
```

例 （源代码: calc.8/src/calculator.js）

```
(function() {

  this.Calculator = (function() {

    function Calculator(scientific) {
      this.scientific = scientific != null ? scientific : false;
    }

    Calculator.prototype.add = function(a, b) {
      return a + b;
    };

    Calculator.prototype.subtract = function(a, b) {
      return a - b;
    };

    Calculator.prototype.multiply = function(a, b) {
      return a * b;
    };

    Calculator.prototype.divide = function(a, b) {
      return a / b;
    };

    return Calculator;

  })();

}).call(this);
```

接下来，我们增加一个测试用例，断言其默认状态下是非科学计算模式。

例（源代码：calc.8/spec/javascripts/calculator_spec.coffee）

```
describe "Calculator", ->

  beforeEach ->
    @calculator = new Calculator()

  it "is not in scientific mode by default", ->
    expect(@calculator.scientific).toBeFalse()

  describe "#add", ->

    it "adds two numbers", ->
      expect(@calculator.add(1, 1)).toEqual 2

  describe "#subtract", ->

    it "subtracts two numbers", ->
      expect(@calculator.subtract(10, 1)).toEqual 9

  describe "#multiply", ->

    it "multiplies two numbers", ->
      expect(@calculator.multiply(5, 4)).toEqual 20

  describe "#divide", ->

    it "divides two numbers", ->
      expect(@calculator.divide(20, 5)).toEqual 4
```

例（源代码：calc.8/spec/javascripts/calculator_spec.js）

```
(function() {

  describe("Calculator", function() {
    beforeEach(function() {
      return this.calculator = new Calculator();
    });
    it("is not in scientific mode by default", function() {
      return expect(this.calculator.scientific).toBeFalse();
    });
    describe("#add", function() {
      return it("adds two numbers", function() {
        return expect(this.calculator.add(1, 1)).toEqual(2);
      });
    });
    describe("#subtract", function() {
      return it("subtracts two numbers", function() {
        return expect(this.calculator.subtract(10, 1)).toEqual(9);
      });
    });
    describe("#multiply", function() {
      return it("multiplies two numbers", function() {
```

```
        return expect(this.calculator.multiply(5, 4)).toEqual(20);
      });
    });
    return describe("#divide", function() {
      return it("divides two numbers", function() {
        return expect(this.calculator.divide(20, 5)).toEqual(4);
      });
    });
  });

}).call(this);

Running Jasmine specs...
....
PASS: 5 tests, 0 failures, 0.021 secs.
```

我们再来书写另一个“describe”代码块，用来描述科学计算模式下的 Calculator 类，我们会在 describe 代码块中增加调用 beforeEach 函数来创建一个科学计算模式下的 Calculator 实例。同时，我们再写一个测试用例来断言它的确是在科学计算模式下。

例（源代码：calc.9/spec/javascripts/calculator_spec.coffee）

```
describe "Calculator", ->

  beforeEach ->
    @calculator = new Calculator()

  it "is not in scientific mode by default", ->
    expect(@calculator.scientific).toBeFalse()

  describe "scientific mode", ->

    beforeEach ->
      @calculator = new Calculator(true)

    it "is in scientific mode when set", ->
      expect(@calculator.scientific).toBeTruth()

  describe "#add", ->

    it "adds two numbers", ->
      expect(@calculator.add(1, 1)).toEqual 2

  describe "#subtract", ->

    it "subtracts two numbers", ->
      expect(@calculator.subtract(10, 1)).toEqual 9

  describe "#multiply", ->

    it "multiplies two numbers", ->
      expect(@calculator.multiply(5, 4)).toEqual 20
```

```
  describe "#divide", ->

    it "divides two numbers", ->
      expect(@calculator.divide(20, 5)).toEqual 4
```

例（源代码: calc.9/spec/javascripts/calculator_spec.js）

```
(function() {

  describe("Calculator", function() {
    beforeEach(function() {
      return this.calculator = new Calculator();
    });
    it("is not in scientific mode by default", function() {
      return expect(this.calculator.scientific).toBeFalse();
    });
    describe("scientific mode", function() {
      beforeEach(function() {
        return this.calculator = new Calculator(true);
      });
      return it("is in scientific mode when set", function() {
        return expect(this.calculator.scientific).toBeTruth();
      });
    });
    describe("#add", function() {
      return it("adds two numbers", function() {
        return expect(this.calculator.add(1, 1)).toEqual(2);
      });
    });
    describe("#subtract", function() {
      return it("subtracts two numbers", function() {
        return expect(this.calculator.subtract(10, 1)).toEqual(9);
      });
    });
    describe("#multiply", function() {
      return it("multiplies two numbers", function() {
        return expect(this.calculator.multiply(5, 4)).toEqual(20);
      });
    });
    return describe("#divide", function() {
      return it("divides two numbers", function() {
        return expect(this.calculator.divide(20, 5)).toEqual(4);
      });
    });
  });

}).call(this);
```

```
Running Jasmine specs...
......
PASS: 6 tests, 0 failures, 0.017 secs.
```

8.6 自定义匹配器

在完成开发 Calculator 类之前，我们可以使用自定义匹配器来对测试计算器是否在科学计算模式下的测试代码进行重构，以使我们的代码更为整洁。Jasmine 为定义自定义匹配器提供了简单、好用的方式。

在 spec/javascripts/helpers 目录下，我们来创建一个名为 to_be_scientific.coffee 的文件。

> **提示：** spec/javascripts/helpers 目录下文件的文件名是什么其实无所谓，不过，应该要确保能准确地描述文件内容。这里我们直接使用要编写的匹配器名字为文件名是一种很好的方式，因为后期需要修改的话，可以很容易找到该文件。

将如下内容添加到刚刚创建的文件中。

例 （源代码：calc.10/spec/javascripts/helpers/to_be_scientific.coffee）

```
beforeEach ->
  @addMatchers
    toBeScientific: ->
      @actual.scientific is true
```

例 （源代码：calc.10/spec/javascripts/helpers/to_be_scientific.js）

```
(function() {

  beforeEach(function() {
    return this.addMatchers({
      toBeScientific: function() {
        return this.actual.scientific === true;
      }
    });
  });

}).call(this);
```

> **提示：** 不需要为每个自定义匹配器创建一个文件。你也可以将它们都定义在一个帮助文件中。不过，我个人倾向于一个匹配器一个文件，这样可以让每个文件中的代码很简短，而且很清楚。你也可以将“一次性”的匹配器写在特定测试用例的“describe”代码块中。

要定义自定义匹配器，首先要做的就是增加 beforeEach 的调用，测试集中每个测试用例执行前都会调用它。在 beforeEach 内部，要调用 Jasmine 内置的函数 addMatchers，顾名思义，该函数就是用来添加自定义匹配器的。它接收一个包含匹配器名字和匹配器内容（函数）的对象作为参数。要谨记，自定义匹配器只能返回 true 或者 false。还记得此前我们提到的，在 Jasmine 中，如果断言返回 true 则表示测试用例通过，反之，则失败吗？这里就是在定义该行为，所以返回 true 或者 false。

在我们的匹配器 toBeScientific 中，要断言 Calculator 实例是否设置了 scientific 标志，并根据该标志返回 true 或者 false。

有了自定义匹配器之后，我们来将测试用例更新为如下形式。

例（源代码：calc.10/spec/javascripts/calculator_spec.coffee）

```
describe "Calculator", ->

  beforeEach ->
    @calculator = new Calculator()

  it "is not in scientific mode by default", ->
    expect(@calculator).not.toBeScientific()

  describe "scientific mode", ->

    beforeEach ->
      @calculator = new Calculator(true)

    it "is in scientific mode when set", ->
      expect(@calculator).toBeScientific()

  describe "#add", ->

    it "adds two numbers", ->
      expect(@calculator.add(1, 1)).toEqual 2

  describe "#subtract", ->

    it "subtracts two numbers", ->
      expect(@calculator.subtract(10, 1)).toEqual 9

  describe "#multiply", ->

    it "multiplies two numbers", ->
      expect(@calculator.multiply(5, 4)).toEqual 20

  describe "#divide", ->

    it "divides two numbers", ->
      expect(@calculator.divide(20, 5)).toEqual 4
```

例（源代码：calc.10/spec/javascripts/calculator_spec.js）

```
(function() {

  describe("Calculator", function() {
    beforeEach(function() {
      return this.calculator = new Calculator();
```

```
    });
    it("is not in scientific mode by default", function() {
      return expect(this.calculator).not.toBeScientific();
    });
    describe("scientific mode", function() {
      beforeEach(function() {
        return this.calculator = new Calculator(true);
      });
      return it("is in scientific mode when set", function() {
        return expect(this.calculator).toBeScientific();
      });
    });
    describe("#add", function() {
      return it("adds two numbers", function() {
        return expect(this.calculator.add(1, 1)).toEqual(2);
      });
    });
    describe("#subtract", function() {
      return it("subtracts two numbers", function() {
        return expect(this.calculator.subtract(10, 1)).toEqual(9);
      });
    });
    describe("#multiply", function() {
      return it("multiplies two numbers", function() {
        return expect(this.calculator.multiply(5, 4)).toEqual(20);
      });
    });
    return describe("#divide", function() {
      return it("divides two numbers", function() {
        return expect(this.calculator.divide(20, 5)).toEqual(4);
      });
    });
  });

}).call(this);
```

代码整洁多了吧？我们的自定义匹配器非常简单，不过，我们也可以很容易地在里面添加更加复杂的逻辑。比方说，如果`Calculator`类有界面，我们就可以在`toBeScientific`匹配器中测试是否设置了`scientific`标志，以及界面上对应的滑块是否切换到了科学技术模式下。

8.7　小结

本章简单介绍了 Jasmine 测试框架。介绍了测试的重要性以及测试驱动开发的好处。我希望通过本章的内容可以让你觉得 TDD 其实很简单并且值得一试。

接下来简要介绍了安装和配置 Jasmine 的方式，特别是如何让 Jasmine 能够在 CoffeeScript

下使用。准备好了 Jasmine 的环境之后，介绍了一个 Jasmine 测试用例。

紧接着，我们为 Calculator 类定义了测试用例，由于还没有实现 Calculator 类，所以一开始所有的测试用例都失败了。随着对该类的实现，测试用例也都一一通过。

最后，我们对测试用例做了多次迭代修改，介绍了如何使用 beforeEach 以及如何定义自定义匹配器。

希望本章介绍 Jasmine 的内容对你有所帮助。另外，还有大量的第三方库可以帮助你更好地编写测试用例，包括可以帮助你测试 UI 元素。可以通过在 GitHub[1]中搜索来找到你需要的库。

结束本章内容前还有最后一件事情：我要你在此向我保证，不论代码是用 CoffeeScript、JavaScript、Java、ColdFusion 写的，还是 Cobalt 写的，你都会为其写测试用例。请伸出你的手宣读如下誓言：

"
我庄严宣誓会对我的代码进行测试。
我并非测试部分，而是测试全部。
我深知不做测试的结果无异于拿起石头砸自己的脚。
做测试并非仅仅为了我自己，更是为了使用我的代码的所有开发者。
我承诺会让其他开发者也做此承诺。
他们如若不从，定会受到惩罚。
测试万岁！
"

恭喜！现在就去测试吧！

① https://github.com/pivotal/jasmine/wiki/Matchers/

第 9 章

Node.js 介绍

前几年，Joyent[①]开发的平台撼动了整个 Web 开发界，该平台叫做 Node.js[②]，通常称为 Node（本章中也会用 Node 这个名字）。

> **提示**：如果你对 Node.js 撼动了整个 Web 开发界持怀疑态度，那么你问问像 LinkedIn[③]这样的公司就明白了。LinkedIn 在 2011 年将它们的 API 从 Ruby on Rails[④]迁移到了 Node 平台。这就是对 Node 最大的认可。

那么究竟什么是 Node，为什么要在一本 CoffeeScript 的书中介绍它呢？本章会做出解答。

9.1 什么是 Node.js

Node 是服务器端 JavaScript 的一种实现，它内部使用了 Google 的 V8[⑤] Javascript 引擎。Node 既是框架也是运行时，同时还可以算是一门语言。正因如此，很多人都很困惑到底 Node 是什么，该什么时候使用 Node 以及为什么要使用 Node。

Node 旨在帮助开发者编写异步的、事件驱动的应用。在 Node 中，每一个请求都是异步的，并且几乎所有的 I/O 都是非阻塞的。正因如此，Node 应用非常高效率，能够处理的“并发”连接的数量也很庞大。

之所以在本书中介绍 Node.js 出于 3 个原因。首先，Node 应用是用 JavaScript 编写的，因此，我们同样可以使用 CoffeeScript 来代替。其次，本书中使用的 `coffee` 命令，就是基于 Node 实现

① http://www.joyentcloud.com/
② http://nodejs.org/
③ http://www.linkedin.com
④ http://www.rubyonrails.org
⑤ http://code.google.com/p/v8/

的，所以，CoffeeScript 和 Node 之间本身就关系密切。最后，Node 自带包管理系统——NPM①。NPM 允许开发者将代码打包发布，让其他开发者作为模块引入使用。这些 NPM 模块非常有用，哪怕不是用来构建 Node 应用；比方说，将它引入到 Cakefile 中来协助完成特定任务。

本章中，我们会构建一个简单的静态 Web 服务器，托管本地目录中的静态文件。除此之外，它还负责将所有 CoffeeScript 资源实时编译后返回给浏览器。毋庸置疑，我们将用 CoffeeScript 来构建此服务器。

9.2 安装 Node

相信你已经很了解我的习惯了，我倾向于让你自己去供应商的网站查看安装指南。因此，要了解如何在你选择的平台上安装 Node，推荐你访问 Node 官网的安装介绍页面：http://nodejs.org/#download。如果你对书中的例子都自己动手实践过，那么你应当已经安装好 Node 了。不过，即使如此，这也是一个更新到最新版本的好机会。

安装好 Node 之后，我们可以采用如下方式来使用 Node 的 REPL：

```
> node
```

然后，我们就可以在 Node REPL 中执行 JavaScript 脚本了，就像这样：

```
node> 1 + 1
```

除了 Node REPL 之外，我们还可以和执行 CoffeeScript 文件一样执行 JavaScript 文件。拿下面这个 JavaScript 文件为例。

例（源代码：example.js）

```
(function() {
  var sayHi;

  sayHi = function(name) {
    if (name == null) name = 'World';
    return console.log("Hello, " + name);
  };

  sayHi('Mark');

}).call(this);
```

我们采用如下方式来执行上述文件：

```
> node example.js
```

① http://npmjs.org/

然后会得到如下输出结果。

输出 （源代码：example.js）

```
Hello, Mark
```

9.3　从这里开始

至此，我们已经安装好Node，并已知道如何用Node来运行JavaScript文件，下面，我们来用Node写一个简单的“Hello，World”服务器。我会先给出实现服务器的代码，随后再对代码做详细解释。

例 （源代码：server.1/server.coffee）

```
http = require('http')

port = 3000
ip = "127.0.0.1"

server = http.createServer (req, res) ->
  data = "Hello World!"
  res.writeHead 200,
    "Content-Type": "text/plain"
    "Content-Length": Buffer.byteLength(data, "utf-8")
  res.write(data, "utf-8")
  res.end()

server.listen(port, ip)

console.log "Server running at http://#{ip}:#{port}/"
```

例 （源代码：server.1/server.js）

```
(function() {
  var http, ip, port, server;

  http = require('http');

  port = 3000;

  ip = "127.0.0.1";

  server = http.createServer(function(req, res) {
    var data;
    data = "Hello World!";
    res.writeHead(200, {
      "Content-Type": "text/plain",
      "Content-Length": Buffer.byteLength(data, "utf-8")
    });
    res.write(data, "utf-8");
    return res.end();
  });
```

```
  server.listen(port, ip);

  console.log("Server running at http://" + ip + ":" + port + "/");

}).call(this);
```

在解释上述代码前，我们先来将服务器启动起来，看看会发生什么：

```
> coffee server.1/server.coffee
```

现在，访问 http://127.0.0.1:3000 就应该能看到欢迎信息了。

来看一个显而易见的问题：为什么这里要用 coffee 命令来执行，而不用我们刚讲的 node 命令呢？原因就在于 node 命令并不知道如何编译、执行 CoffeeScript 脚本。如果用 node 命令来执行 CoffeeScript 文件，会报错。幸运的是，coffee 命令是基于 Node 的，所以，我们可以用 coffee 命令直接执行。

> **提示：**如果真想用 node 命令来执行的话，可以将 CoffeeScript 文件先编译成 JavaScript 文件，然后再执行。不过，有直接的 coffee 命令不用，为何还要多此一举呢？

我们来剖析上述这段代码，看看究竟是如何构建起这样一个 Web 服务器的。

首先，我们需要导入 Node 内置的 http 模块。该模块完成了所有我们需要的处理和接收 HTTP 请求的工作。通过 Node 提供的 require 函数就可以导入该模块。接下来，我们创建了一些服务器所需的表示端口和 IP 地址的变量。

然后，我们通过调用 http 模块提供的 createServer 函数，创建了 http 服务器。我们给该函数传递了一个回调函数，用于处理新进来的请求。这里最酷的地方就在于，只有当请求进来的时候，才会触发我们传入的回调函数；其余时间，服务器都处于监听等待的状态，这种时候只会占用很少的资源。我们先跳过回调函数的具体内容，来看看其余部分的代码。剩下的代码主要实现服务器监听指定的 IP 机器上的指定端口。

在我们的回调函数中，接收两个参数。第一个参数是请求对象，第二个是要发送回去的响应对象。req 参数（也就是请求对象）中包括大量有用的数据——请求页面信息、浏览器类型、查询字段，等等。

> **提示：**我建议在回调函数中加入 console.log 语句，用来打印出该请求对象来一探究竟。这是一种快速而又简单的查看每个请求的方式。

当服务器接收到请求时，会调用回调函数，我们可以通过传递进来的请求和对象来处理响应。上述例子中，我们首先做的是定义了一个叫 data 的变量，该变量存放了要响应的数据——“Hello World!”

现在我们知道要说什么了，就可以将响应回写到客户端。首先，我们必须填充响应码和响应头；这里我们通过 response 对象上的 writeHead 函数来实现。writeHead 函数的第一个参数是 HTTP 的状态码。上述例子中因为一切正常，所以返回 200。第二个参数是包含 HTTP 响应头信息的对象。上述例子中，我们定义了两个响应头。第一个是 Content-Type，因为我们要返回的是只是文本数据，并非二进制或者 HTML 数据，所以将该字段值设置为 text/plain。第二个是 Content-Length。该字段告诉客户端，返回的数据大小是多少，以字节为单位。为此，Node 提供了简单好用的函数 Buffer.byteLength。只需简单地将数据传递给该函数就能知道数据长度为多少个字节。

在将状态码和响应头信息返回给客户端后，我们可以用 write 函数将数据回写给客户端了。

在将数据回写给客户端之后，最后，我们调用 end 函数，顾名思义，它是用来结束本次响应的。至此，我们就成功构建了第一个简单的 Node 服务器。恭喜恭喜！

9.4　流化响应

编写 Node 应用时，有一点很重要：在处理请求，往响应体写数据时，并非只能调用 write 函数一次。你可以在调用 end 函数前，调用任意多次 write 函数。这为编写流式 API 提供了很好的方式。让我们来一探究竟。

我们来写一个简单的服务器，打印一段电影《闪灵》（*The Shining*）[1]中的经典台词——“All work and no play makes Jack a dull boy”。并在 30 秒内持续不断地打印该信息。

例（源代码：streaming/server.coffee）

```
http = require('http')

port = 3000
ip = "127.0.0.1"

server = http.createServer (req, res) ->
  res.writeHead 200,
    "Content-Type": "text/plain"
  setInterval ->
    res.write("All work and no play makes Jack a dull boy. ")
  , 10
  setTimeout ->
    res.end()
  , 30000

server.listen(port, ip)

console.log "Server running at http://#{ip}:#{port}/"
```

例（源代码：streaming/server.js）

```
(function() {
  var http, ip, port, server;
```

[1] http://www.imdb.com/title/tt0081505/

```
  http = require('http');

  port = 3000;

  ip = "127.0.0.1";

  server = http.createServer(function(req, res) {
    res.writeHead(200, {
      "Content-Type": "text/plain"
    });
    setInterval(function() {
      return res.write("All work and no play makes Jack a dull boy. ");
    }, 10);
    return setTimeout(function() {
      return res.end();
    }, 30000);
  });

  server.listen(port, ip);

  console.log("Server running at http://" + ip + ":" + port + "/");

}).call(this);
```

```
> coffee streaming/server.coffee
```

打开浏览器，访问 `http://127.0.0.1:3000`，就会看到页面上这些信息持续不断地输出在页面上，并在 30 秒后结束。如果是非流式的话，我们会在 30 秒内看到一张空白页面，这个时候我们甚至会怀疑是不是服务器出错了，直到 30 秒后我们会突然一下子看到所有的输出信息。

在回调函数中，我们设置了 `setInterval` 循环，每 10 毫秒向响应体写入信息。之后，通过 `setTimeout` 来实现在 30 秒后完成响应。

总的来说，用 Node 构建一个流式的服务器非常容易。如果想要看一些真正酷的东西，在浏览器中打开多个窗口，分别访问该服务器。你会发现，Node 很容易地将这些请求分开处理并流式响应给浏览器，而这一切不需要任何额外的代码来处理。

9.5 构建 CoffeeScript 服务器

我们来构建一个有点意义的服务器。需求很简单。服务器查看进来的请求，然后根据情况做以下三件事情之一。如果是请求一个 JavaScript 文件，则查找 `src` 目录中对应的 CoffeeScript 文件，编译该文件，并将编译后的 JavaScript 文件返回。如果请求的文件或者路径不是 JavaScript 文件，则返回 `public` 目录中找到对应的文件。最后，如果在 `src` 目录和 `public` 目录中都没有找到对应的文件，则返回 404 页面。

了解了需求，在实现前，你应当知道这次的代码量会很大，我们会进行迭代开发。也正因如此，过程中我不会给你展示编译后的 JavaScript 文件，直到最终完成后，再将其展示出来。相信我，你一定会感谢我这种做法的。好了，现在就开始吧。

进一步思考，实现该服务器，回调函数中的代码肯定很多，所以我们将这部分代码抽离出来放到一个类中来单独处理。假设我们已经实现了该类，现在来写回调函数的内容，这样，恰好也能知道我们需要的该类的功能。

例（源代码：server.2/server.coffee）

```
http = require('http')

port = 3000
ip = "127.0.0.1"

server = http.createServer (req, res) ->
  app = new Application(req, res)
  app.process()

server.listen(port, ip)

console.log "Server running at http://#{ip}:#{port}/"
```

> **提示**：不将所有代码都放在回调函数中，还有另外一个好处：容易测试。如果我们创建了一些类来处理请求，那么我们就可以采用第 8 章学到的来对每个文件进行单独测试，确保它们工作正常，而不需要为了测试还要在后台启动一个服务器。

上述代码是没法运行起来的，因为我们还没有真正实现 Application 类，下面我们就来实现它。

例（源代码：server.3/server.coffee）

```
http = require('http')

class Application

  constructor: (@req, @res) ->

  process: ->


port = 3000
ip = "127.0.0.1"

server = http.createServer (req, res) ->
  app = new Application(req, res)
  app.process()

server.listen(port, ip)

console.log "Server running at http://#{ip}:#{port}/"
```

Application 类有了很漂亮的“骨架”。下面我们再来让它有点“肉感”。

例（源代码：server.4/server.coffee）

```
http = require('http')
url = require('url')

class Application

  constructor: (@req, @res) ->
    @pathInfo = url.parse(@req.url, true)

  process: ->
    if /^\/javascripts\//.test @pathInfo.pathname
      new JavaScriptProcessor(@req, @res, @pathInfo).process()
    else
      new PublicProcessor(@req, @res, @pathInfo).process()

port = 3000
ip = "127.0.0.1"

server = http.createServer (req, res) ->
  app = new Application(req, res)
  app.process()

server.listen(port, ip)

console.log "Server running at http://#{ip}:#{port}/"
```

首先，我们需要引入 Node 内置的 url 模块。该模块提供了一些帮助函数，可以用来将请求路径分割为几部分。我们将这部分的处理放在了 Application 类的构造函数中。

process 函数中就比较有意思了。我们首先需要检查请求类型，然后做出相应的处理。根据需求，我们只需处理两类请求：JavaScript 文件和其他。所以，我们写两个类，一个负责处理所有的 JavaScript 文件，另外一个负责所有 public 目录中的文件。通过一段简单的正则表达式就可以检查请求类型，然后，调用对应的类来处理该请求类型就可以了。

> **提示**：如果在实际应用中，我会强烈建议将这些类的定义放在单独的文件中，不过，这里我们为了简单起见，就定义在一个文件中了。

下面，我们来定义几个不同处理器所需的函数。我们大概能猜到 JavaScriptProcessor 和 PublicProcessor 都有共同的功能，因此为它们创建一个父类，定义这些共同的功能。下面就来定义 Processor 父类、JavaScriptProcessor 类和 PublicProcessor 类。

例（源代码：server.5/server.coffee）

```
http = require('http')
url = require('url')

class Application
```

```
  constructor: (@req, @res) ->
    @pathInfo = url.parse(@req.url, true)

  process: ->
    if /^\/javascripts\//.test @pathInfo.pathname
      new JavaScriptProcessor(@req, @res, @pathInfo).process()
    else
      new PublicProcessor(@req, @res, @pathInfo).process()

class Processor

  constructor: (@req, @res, @pathInfo) ->

  contentType: ->
    throw new Error("must be implemented!")

  process: ->
    throw new Error("must be implemented!")

  pathname: ->

  write: (data, status = 200, headers = {}) ->

class JavaScriptProcessor extends Processor

  contentType: ->

  process: ->

class PublicProcessor extends Processor

  contentType: ->
  process: ->

port = 3000
ip = "127.0.0.1"

server = http.createServer (req, res) ->
  app = new Application(req, res)
  app.process()

server.listen(port, ip)

console.log "Server running at http://#{ip}:#{port}/"
```

先来看看我们的 Processor 类，它有 5 个函数。第一个是构造（constructor）函数。构造函数用来获取请求对象、响应对象以及包含请求路径详细信息的对象。

后面两个函数 contentType 和 process，需要子类来做具体实现。contentType 函数返回文件正确的类型给响应头。process 函数负责最繁重的任务：从 public 目录中读取文件或者编译请求的 CoffeeScript 文件。

接下来的 pathname 函数用来计算要在本地目录中查找的文件名。子类不必非得实现该函数，因为我们可以直接采用请求路径作为默认值来使用，下面会看到。

最后一个 write 函数负责写响应信息，包括响应状态、响应头以及响应体信息。如上述代码所示，我们设置了默认的状态码为 200，默认的响应头为空。后续在子类的实现中，我们会设置更好的默认响应头。

在实现 JavaScriptProcessor 类和 PublicProcessor 类之前，我们先来完成父类 Processor 的实现。

例（源代码：server.6/server.coffee）

```
http = require('http')
url = require('url')

class Application

  constructor: (@req, @res) ->
    @pathInfo = url.parse(@req.url, true)

  process: ->
    if /^\/javascripts\//.test @pathInfo.pathname
      new JavaScriptProcessor(@req, @res, @pathInfo).process()
    else
      new PublicProcessor(@req, @res, @pathInfo).process()

class Processor

  constructor: (@req, @res, @pathInfo) ->

  contentType: ->
    throw new Error("must be implemented!")

  process: ->
    throw new Error("must be implemented!")

  pathname: ->
    @pathInfo.pathname

  write: (data, status = 200, headers = {}) ->
    headers["Content-Type"] ||= @contentType()
    headers["Content-Length"] ||= Buffer.byteLength(data, "utf-8")
    @res.writeHead(status, headers)
    @res.write(data, "utf-8")
    @res.end()

class JavaScriptProcessor extends Processor

  contentType: ->

  process: ->
```

```
class PublicProcessor extends Processor

  contentType: ->

  process: ->

port = 3000
ip = "127.0.0.1"

server = http.createServer (req, res) ->
  app = new Application(req, res)
  app.process()
server.listen(port, ip)

console.log "Server running at http://#{ip}:#{port}/"
```

我们需要做的就是将客户端请求对象中的路径设置为 pathname 函数默认的返回值。另外，write 函数的实现也很清楚，和此前讨论的并无二异。

下面我们来实现那两个更加有趣的处理器，先来实现 JavaScriptProcessor 类。要实现该类，我们需要重写父类中的 3 个函数，并且还需要 2 个 Node 模块，如下所示。

例（源代码：server.7/server.coffee）

```
http = require('http')
url = require('url')
fs = require('fs')
CoffeeScript = require('coffee-script')

class Application

  constructor: (@req, @res) ->
    @pathInfo = url.parse(@req.url, true)

  process: ->
    if /^\/javascripts\//.test @pathInfo.pathname
      new JavaScriptProcessor(@req, @res, @pathInfo).process()
    else
      new PublicProcessor(@req, @res, @pathInfo).process()

class Processor

  constructor: (@req, @res, @pathInfo) ->

  contentType: ->
    throw new Error("must be implemented!")

  process: ->
    throw new Error("must be implemented!")

  pathname: ->
```

```
      @pathInfo.pathname

    write: (data, status = 200, headers = {}) ->
      headers["Content-Type"] ||= @contentType()
      headers["Content-Length"] ||= Buffer.byteLength(data, "utf-8")
      @res.writeHead(status, headers)
      @res.write(data, "utf-8")
      @res.end()

  class JavaScriptProcessor extends Processor

    contentType: ->
      "application/x-javascript"

    pathname: ->
      file = (/\/javascripts\/(.+)\.js/.exec(@pathInfo.pathname))[1]
      return "#{file}.coffee"

    process: ->
      fs.readFile "src/#{@pathname()}", "utf-8", (err, data) =>
        if err?
          @write("", 404)
        else
          @write(CoffeeScript.compile(data))

  class PublicProcessor extends Processor

    contentType: ->

    process: ->

  port = 3000
  ip = "127.0.0.1"

  server = http.createServer (req, res) ->
    app = new Application(req, res)
    app.process()

  server.listen(port, ip)

  console.log "Server running at http://#{ip}:#{port}/"
```

首先需要引入Node内置的`fs`模块。这个名字有点含糊的模块提供了对文件系统的操作。之所以需要用到此模块，是为了能够从磁盘上的文件系统中读取到我们的CoffeeScript文件，并将其编译。同样的，在`PublicProcessor`类中也需要用到此模块来读取`public`目录中的文件。

另外一个需要引入的是`coffee-script`模块。是不是有点混乱，怎么还需要这个模块？事实上，我们是用它来将请求的CoffeeScript文件编译成JavaScript文件。

> **提示**：`coffee-script`模块并非仅提供了`compile`函数，只是，我们在这里只用到此功能而已。我强烈建议去深入挖掘下，看看它还能做什么。

引入了这两个模块后，我们就可以开始实现 JavaScriptProcessor 类必要的函数了。其中，实现 contentType 函数非常简单，直接返回 JavaScript 内容类型——application/x-javascript 即可。

> **提示**：如果用上述内容类型有问题，可能是你使用了某些版本的 IE 浏览器。尝试将其改为 text/javascript;应该可以解决问题。

相对而言，实现 pathname 函数就不像实现 contentType 函数那么容易了，不过，也不算很麻烦。我们可以使用正则表达式来去除请求路径中的 javascripts/和.js，仅留下中间部分。比方说，如果请求的路径为 javascripts/foo.js，那么，通过正则表达式移除后就只剩 foo。然后，我们只要将这部分加上.coffee 后缀即可。

最后，我们需要实现 process 函数。从某种角度来看，process 函数相比我们刚刚实现的 pathname 函数来说，更直观。我们通过 sf 模块的 readFile 函数来从 src 目录中读取 CoffeeScript 文件。readFile 函数随后会执行我们传递进去的回调函数。

传递给 readFile 的回调函数有两个参数。第一个参数是个对象，表示读取文件操作过程中可能遇到的错误，比方说：文件不存在。第二个参数是读取的文件的内容。

在这个回调函数中，我们用第 3 章中学到的有关存在的操作符？来检查是否有错误，如果有错误，就调用父类的 write 函数，并将两个参数分别设置为空内容以及 404 状态码。如果没有错误，则将从磁盘中读取到的 CoffeeScript 文件内容交给 coffee-script 模块的 compile 函数去编译，然后，将编译后的 JavaScript 代码传递给父类的 write 函数，将其回写给客户端。不错吧？

最后剩下的就是实现 PublicProcessor 类中同样的这 3 个函数，这样我们就完成了我们的服务器。

例 （源代码：final/server.coffee）

```
http = require('http')
url = require('url')
fs = require('fs')
CoffeeScript = require('coffee-script')

class Application

  constructor: (@req, @res) ->
    @pathInfo = url.parse(@req.url, true)

  process: ->
    if /^\/javascripts\//.test @pathInfo.pathname
      new JavaScriptProcessor(@req, @res, @pathInfo).process()
    else
      new PublicProcessor(@req, @res, @pathInfo).process()

class Processor

  constructor: (@req, @res, @pathInfo) ->
```

```
  contentType: ->
    throw new Error("must be implemented!")

  process: ->
    throw new Error("must be implemented!")

  pathname: ->
    @pathInfo.pathname

  write: (data, status = 200, headers = {}) ->
    headers["Content-Type"] ||= @contentType()
    headers["Content-Length"] ||= Buffer.byteLength(data, "utf-8")
    @res.writeHead(status, headers)
    @res.write(data, "utf-8")
    @res.end()

class JavaScriptProcessor extends Processor

  contentType: ->
    "application/x-javascript"

  pathname: ->
    file = (/\/javascripts\/(.+)\.js/.exec(@pathInfo.pathname))[1]
    return "#{file}.coffee"
  process: ->
    fs.readFile "src/#{@pathname()}", "utf-8", (err, data) =>
      if err?
        @write("", 404)
      else
        @write(CoffeeScript.compile(data))

class PublicProcessor extends Processor

  contentType: ->
    ext = (/\.(.+)$/.exec(@pathname()))[1].toLowerCase()
    switch ext
      when "png", "jpg", "jpeg", "gif"
        "image/#{ext}"
      when "css"
        "text/css"
      else
        "text/html"

  process: ->
    fs.readFile "public/#{@pathname()}", "utf-8", (err, data) =>
      if err?
        @write("Oops! We couldn't find the page you were looking for.", 404)
      else
        @write(data)

  pathname: ->
```

```
    unless @_pathname
      if @pathInfo.pathname is "/" or @pathInfo.pathname is ""
        @pathInfo.pathname = "index"
      unless /\..+$/.test @pathInfo.pathname
        @pathInfo.pathname += ".html"
      @_pathname = @pathInfo.pathname
    return @_pathname

port = 3000
ip = "127.0.0.1"

server = http.createServer (req, res) ->
  app = new Application(req, res)
  app.process()

server.listen(port, ip)

console.log "Server running at http://#{ip}:#{port}/"
```

例（源代码: final/server.js）

```
(function() {
  var Application, CoffeeScript, JavaScriptProcessor, Processor, PublicProcessor, fs,
➥http, ip, port, server, url,
  __hasProp = Object.prototype.hasOwnProperty,
  __extends = function(child, parent) { for (var key in parent) { if
(__hasProp.call(parent, key)) child[key] = parent[key]; } function ctor()
{ this.constructor = child; } ctor.prototype = parent.prototype; child.prototype =
new ctor; child.__super__ = parent.prototype; return child; };

  http = require('http');

  url = require('url');

  fs = require('fs');

  CoffeeScript = require('coffee-script');

  Application = (function() {

    function Application(req, res) {
      this.req = req;
      this.res = res;
      this.pathInfo = url.parse(this.req.url, true);
    }

    Application.prototype.process = function() {
      if (/^\/javascripts\//.test(this.pathInfo.pathname)) {
        return new JavaScriptProcessor(this.req, this.res, this.pathInfo).process();
      } else {
        return new PublicProcessor(this.req, this.res, this.pathInfo).process();
      }
```

```
  };

  return Application;

})();

Processor = (function() {
  function Processor(req, res, pathInfo) {
    this.req = req;
    this.res = res;
    this.pathInfo = pathInfo;
  }

  Processor.prototype.contentType = function() {
    throw new Error("must be implemented!");
  };

  Processor.prototype.process = function() {
    throw new Error("must be implemented!");
  };

  Processor.prototype.pathname = function() {
    return this.pathInfo.pathname;
  };

  Processor.prototype.write = function(data, status, headers) {
    if (status == null) status = 200;
    if (headers == null) headers = {};
    headers["Content-Type"] || (headers["Content-Type"] = this.contentType());
    headers["Content-Length"] || (headers["Content-Length"] =
➥Buffer.byteLength(data, "utf-8"));

    this.res.writeHead(status, headers);
    this.res.write(data, "utf-8");
    return this.res.end();
  };

  return Processor;

})();

JavaScriptProcessor = (function(_super) {

  __extends(JavaScriptProcessor, _super);

  function JavaScriptProcessor() {
    JavaScriptProcessor.__super__.constructor.apply(this, arguments);
  }

  JavaScriptProcessor.prototype.contentType = function() {
```

```
    return "application/x-javascript";
  };

  JavaScriptProcessor.prototype.pathname = function() {
    var file;
    file = (/\/javascripts\/(.+)\.js/.exec(this.pathInfo.pathname))[1];
    return "" + file + ".coffee";
  };

  JavaScriptProcessor.prototype.process = function() {
    var _this = this;
    return fs.readFile("src/" + (this.pathname()), "utf-8", function(err, data) {
      if (err != null) {
        return _this.write("", 404);
      } else {
        return _this.write(CoffeeScript.compile(data));
      }
    });
  };

  return JavaScriptProcessor;

})(Processor);

PublicProcessor = (function(_super) {

  __extends(PublicProcessor, _super);

  function PublicProcessor() {
    PublicProcessor.__super__.constructor.apply(this, arguments);
  }

  PublicProcessor.prototype.contentType = function() {
    var ext;
    ext = (/\.(.+)$/.exec(this.pathname()))[1].toLowerCase();
    switch (ext) {
      case "png":
      case "jpg":
      case "jpeg":
      case "gif":
        return "image/" + ext;
      case "css":
        return "text/css";
      default:
        return "text/html";
    }
  };
  PublicProcessor.prototype.process = function() {
    var _this = this;
    return fs.readFile("public/" + (this.pathname()), "utf-8", function(err, data) {
```

```
        if (err != null) {
          return _this.write("Oops! We couldn't find the page you were looking for.",
➥404);
        } else {
          return _this.write(data);
        }
      });
    };

    PublicProcessor.prototype.pathname = function() {
      if (!this._pathname) {
        if (this.pathInfo.pathname === "/" || this.pathInfo.pathname === "") {
          this.pathInfo.pathname = "index";
        }
        if (!/\..+$/.test(this.pathInfo.pathname)) {
          this.pathInfo.pathname += ".html";
        }
        this._pathname = this.pathInfo.pathname;
      }
      return this._pathname;
    };

    return PublicProcessor;

  })(Processor);

  port = 3000;

  ip = "127.0.0.1";

  server = http.createServer(function(req, res) {
    var app;
    app = new Application(req, res);
    return app.process();
  });

  server.listen(port, ip);

  console.log("Server running at http://" + ip + ":" + port + "/");

}).call(this);
```

PublicProcessor 中 process 函数的实现和此前实现的 JavaScriptProcessor 中的没多大区别。最大的两个区别就是：当遇到错误时，不是传递空字符串，而是传递了一段错误信息；还有就是，发送文件内容前不再需要编译了。

contentType 方法看上去可能有点儿困惑，因为我们除了想要处理 HTML 文件还想要能处理一些图片。首先，我们需要通过一段正则表达式来获取请求路径中的文件扩展名。然后，通过第 3 章中介绍的 switch 语句，为不同的扩展名对应相应的文件类型。

> **提示**：此刻，为了避免误会，我要指出，我们实现的服务器仅仅是用于教学目的，不能投入到实际产品中。因为这里面还存在很多的问题，我们没有处理。比方说：对错误进行正确的处理，以及内容类型的检测。所以，如果你将其用到实际产品中并出现了问题，请不要发邮件责问我。

最后是实现 pathname 函数。这可能是整个例子中最复杂的函数，因为我们要处理三种情况。第一种是处理完整路径的情况，如：index.html 或者“images/foo.png”。这种情况下路径是完好的，所以直接返回即可。

第二种是路径中不包含扩展名的，如 index 或者 users/1。由于这类路径缺少文件扩展名，所以要为它们添加一个默认的扩展名.html，将其转化为 index.html 以及 users/1.html。

最后一种是默认的路径/。需要将其转化为 index.html，以便能在 public 目录中查找它。

你应当注意到了，我将最后的结果缓存到了@_pathname 变量中，这样做的目的是为了只让其在首次被调用的时候执行一次，之后直接返回结果即可。

9.6 验收我们的服务器

构建好服务器后，怎么用呢？我们先来创建一个 src 目录，同时在该目录下创建一个 application.coffee 文件，内容如下。

例（源代码：final/src/application.coffee）

```
$ ->
  $("body").html("Hello from jQuery and Node!!")
```

例（源代码：final/src/application.js）

```
(function() {

  $(function() {
    return $("body").html("Hello from jQuery and Node!!");
  });

}).call(this);
```

如果你对 jQuery[①]不熟悉，我来解释下上述代码。上述代码用来在 DOM[②]载入后，将 HTML 文件中的 body 内容替换为“Hello from jQuery and Node!!”。

下面，我们在 public 目录中创建 index.html 文件。

例（源代码：final/public/index.html）

```
<!DOCTYPE html>
<html>
```

① http://jquery.com/
② http://en.wikipedia.org/wiki/Document_Object_Model

```
  <head>
    <title>Welcome to Node.js</title>
    <script src="http://ajax.googleapis.com/ajax/libs/jquery/1.7.1/jquery.min.js"
➥type="text/javascript"></script>
    <script src="/javascripts/application.js" type="text/javascript"></script>
  </head>
  <body>
    Hello from Node!!
  </body>
</html>
```

上述 HTML 文件非常简单。在 HTML 头部，我们引入两个 JavaScript 文件，分别是 jQuery 以及我们服务器上的/javascripts/application.js 文件，如果没问题的话，该文件会映射到 src/application.coffee 文件。在 HTML 的正文中，我们打印出“Hello from Node!!”。

最后，启动服务器并访问 http://127.0.0.1:3000。

```
> coffee finale/server.coffee
```

你应当能看到“Hello from jQuery and Node!!”的欢迎语而不再是“Hello from Node!!”。

9.7　小结

本章对 Node.js 做了简单的介绍。我们介绍了什么是 Node.js 以及它的应用场景，还介绍了如何用 Node 来简单地实现一个流式服务器。

我们还构建了一个有趣的服务器，用来自动对请求文件进行对应 CoffeeScript 文件的自动编译。

Node.js 是一个用于构建真正可扩展的、事件驱动的应用的平台。除了构建 HTTP 服务器外，还可以构建处理套接字的应用，比方说 Telnet，或者聊天应用都可以。另外，除了流式响应外，还可将此逆转。流化客户端请求。比方说，处理上传文件就是其中一个例子！有了 Node.js，这一切皆有可能。

有大量优秀的深入介绍 Node 的教程和截屏视频，强烈建议你把它们找出来，多了解一下 Node 都能做些什么。

第 10 章

示例：待办事宜列表第 1 部分（服务器端）

本书第一部分介绍了 CoffeeScript 的优缺点。我们介绍了每一个细节：该语言如何工作、如何映射到 JavaScript 代码、如何帮助编写更加严谨更加整洁的应用。在后面的几章内容中，我们看了几个用 CoffeeScript 投建的项目。现在，在本书最后的三章内容中，我会介绍如何将 CoffeeScript 投入到实际应用中。

对我个人而言，最好的学习方式就是实践。看些有关库或特性的简单例子最多只能教会我用那个库能做什么，而如果我自己亲自实践下，那结果就截然不同了。我能看到它如何能够应用到我日常的开发中去。简而言之，实践是一种很好的方式。这也是为什么我们要在最后三章中来实践开发一个应用的原因。

在最后三章中，我们要开发的应用是经典的待办事宜列表应用。它是最常见的展示案例，相信大家都写过。之所以采用它作为示例，是因为大家都了解该应用的功能以及它是如何工作的。同时，它规模适中，既能帮助我们练习学到的库和工具，又能让我们在相对短的时间内完成一个完整的项目。

本章构建应用的服务器端（后端）部分。我们需要服务器来托管待办事宜列表的页面、资源文件，最为重要的是，托管待办事宜应用本身，可以将数据持久化起来。

在随后的两章中，我们会构建应用的客户端（前端）部分。我们需要一种展示待办事宜页面、创建新待办事宜、更新已有待办事宜以及删除待办事宜的方式。

最后，我想到目前为止你已经看了很多的 JavaScript 代码，所以接下来我们不再展示 CoffeeScript 编译后的 JavaScript 代码。如果你真的要看的话，可以自己去编译或者访问本书的 GitHub 项目[1]进行查看。

---

① https://github.com/markbates/Programming-In-CoffeeScript

10.1 安装并设置 Express

构建最好的待办事宜应用首先要做的就是挑选一个 Web 服务器/框架来处理后端请求。为此，我们选择最流行的 Node 框架——Express①。Express 的安装、设置和使用都非常简单，并且它有非常活跃的开发者社区，所以它是个很好的选择。

开始，我们先找一个干净的目录，创建一个 app.coffee 文件。目前就让该文件为空好了，我们先来安装 Express。

Express 是以 Node 包管理器（Node Package Manager，NPM）②的模块形式发布的。在第 9 章中，我们简单地提到过 NPM。NPM 的模块是打包好并发布的代码，可以被安装以及在 Node 应用中引入。如果你已经安装好 Node 了，那么 NPM 也就有了。

通过下面一行简单的命令就可以安装 Express：

```
> npm install express
```

然后，会看到类似下面这样的输出信息：

```
express@2.5.2 ./node_modules/express
— mime@1.2.4
— mkdirp@0.0.7
— qs@0.4.0
— con 2nect@1.8.3
```

现在你应该能在应用目录下看到一个叫 node_modules 的文件夹。NPM 会将所有安装的模块放到该文件夹中，后面我们还会在该目录下安装更多的模块的。

> **提示**：通过设置-g 标志，可以全局安装 NPM 模块：npm install -g express。我倾向于在项目中安装模块，这样可以将模块也嵌入到 SCM 中，开发起来更方便，避免了很多潜在的版本冲突的问题。

安装好 Express 之后，我们来填充此前创建的 app.coffee 文件的内容。

例（源代码：app.1/app.coffee）

```
# Setup Express.js:
global.express = require('express')
global.app = app = express.createServer()
require("#{__dirname}/src/configuration")

# Set up a routing for our homepage:
```

① http://expressjs.com/
② http://npmjs.org/

```
require("#{__dirname}/src/controllers/home_controller")

# Start server:
app.listen(3000)
console.log("Express server listening on port %d in %s mode", app.address().port,
➥app.settings.env)
```

我们来对上述代码做下说明。首先，我们引入了 express 模块。然后，调用了 createServer 函数，顾名思义，该函数会创建一个服务器。随后，我们将创建出来的服务器赋值给两个变量，一个本地变量 app 和一个全局变量，后者也叫 app。

> **提示：**在 Node 中，global 对象是对整个应用全局作用域的引用，和浏览器中的 window 对象类似。通过将对象绑定到 global 对象上，可以在应用中任何位置引用到。

接下来，需要配置 Express。我们将这部分放到另外一个 src/configuration.coffee 文件中。

例（源代码：app.1/src/configuration.coffee）

```
# Configure Express.js:
app.configure ->
  app.use(express.bodyParser())
  app.use(express.methodOverride())
  app.use(express.cookieParser())
  app.use(express.session(secret: 'd19e19fd62f62a216ecf7d7b1de434ad'))
  app.use(app.router)
  app.use(express.static(__dirname + '../public'))
  app.use(express.errorHandler(dumpExceptions: true, showStack: true))
```

我们配置 app 时使用了一些基础的中间件，例如：会话（session）、cookie、路由、错误处理以及 HTTP 消息体的解析。同时还配置了静态资源的路径，告诉服务器在哪里可以找到像图片、HTML 这样的静态文件。

> **提示：**在 Node 中，变量__dirname 会返回当前执行文件所在的目录。此变量对构建其他文件的路径非常有用。

接下来，我们为首页设置路由规则。我们将这部分功能工作放在 src/controllers/home_controller.coffee 中。app.coffee 通过 Node 的 require 函数引入了 home_controller.coffee 文件，这个文件能让我们在应用中载入其他文件。

例（源代码：app.1/src/controllers/home_controller.coffee）

```
# Set up a routing for our homepage:
app.get '/', (req, res) ->
  res.send "Hello, World!"
```

在具有全局作用域的 app 对象上，需要映射对首页的路由规则。我们可以使用 app 对象上的 get 函数来实现。app 对象上有 4 个对应了 HTTP 行为的函数——get、post、put 和 delete。因为这里希望用户通过 GET 请求就能访问首页，所以我们配置成上述形式。

除了接收要路由的 URL——/之外，get 函数还接收一个回调函数，用来处理匹配指定 URL 的请求。该回调函数接收一个请求对象——req 和一个响应对象——res。上述例子中，我们返回 Hello, World!给客户端。

默认路由配置好之后，最后在 app.coffee 中要做的就是告诉服务器要监听哪个端口了。

下面，我们来将服务器启动起来！

```
> coffee app.coffee
```

接下来，访问 http://localhost:3000，你应当就能看到页面上的“Hello, World!”欢迎语了。

恭喜，你现在已经将一个 Express 应用运行起来了！

应用跑起来之后，我们来给它增加一个模板引擎，这样就不用将所有的 HTML 都放在 home_controller.coffee 文件中了，直接以单独文件形式存在即可。为此需要以 NPM 模块的形式安装 ejs，这样就可以在模板文件中嵌入 JavaScript 代码：

```
> npm install ejs
```

你会看到类似下面这样的输出：

```
ejs@0.6.1 ./node_modules/ejs
```

> **提示**：尽管我们也可以直接简单地托管静态 HTML 文件，但用了 ejs 模板引擎后，还可以很容易地在 HTML 页面中嵌入动态内容。这非常有用。

接下来，我们要告诉 Express 去哪里查找模板文件，而且要用 ejs 模块来处理这些模板。怎么做呢？我们在 src/configuration.coffee 文件最后加上两行代码。

例 （源代码：app.2/src/configuration.coffee）

```
# Configure Express.js:
app.configure ->
  app.use(express.bodyParser())
  app.use(express.methodOverride())
  app.use(express.cookieParser())
  app.use(express.session(secret: 'd19e19fd62f62a216ecf7d7b1de434ad'))
  app.use(app.router)
  app.use(express.static(__dirname + '../public'))
  app.use(express.errorHandler(dumpExceptions: true, showStack: true))
  app.set('views', "#{__dirname}/views")
  app.set('view engine', 'ejs')
```

我们来创建一个新的 src/views/index.ejs 文件。当用户通过我们定义好的路径访问时，会看到此页面，此页面作为应用的首页使用。我们将此页面做得相当简单——同样也是显示"Hello, World!"，不过除此之外，我们还加入了一段动态内容，来验证模板引擎是否可用。

例 （源代码：app.2/src/views/index.ejs）

```
<!DOCTYPE html>
<html>
  <head>
    <title>Todos</title>
  </head>
  <body>
    <h1>Hello, World!</h1>
    <h2>The date/time is: <%= new Date() %></h2>
  </body>
</html>
```

最后，需要更新 home_controller.coffee 文件以使用新的 index.ejs 文件。

例 （源代码：app.2/src/controllers/home_controller.coffee）

```
# Set up a routing for our homepage:
app.get '/', (req, res) ->
  res.render 'index', layout: false
```

我们要做的就是，使用 render 函数替换原来的 sender 函数，将要用的模板名字传进去即可。同时，我们还告诉模板引擎无需去查找布局（layout）模板。

> **提示**：有很重要的一点要注意：所有 ejs 模板中的动态内容都要用 JavaScript 来写，不能用 CoffeeScript。这的确有点儿让人失望。当然，还有一些比较好的基于 CoffeeScript 的模板引擎，不过配置起来相对比较麻烦。这里我推荐 Eco[①]和 CoffeeKup[②]这两款引擎。

重启服务器访问后，你应当能够看到新首页，其中包括当前日期和时间信息。

让我们进入下一步：建立数据库。

10.2 使用 Mongoose 建立 MongoDB 数据库

我们需要将应用的数据持久化下来，因此打算使用 MongoDB[③]。MongoDB 也简称为 Mongo，是一个流行的 NoSQL 文档型数据库。它和关系型数据库一样可以存储和获取数据，不过它更为灵活，不需要传统的模式（schema），这意味着不需要写一大堆创建表的脚本就可以直接使用它。这就是我们选择 MongoDB 的原因，可以直接将待办事宜的数据"丢"到数据库里就可以了。

---

① https://github.com/sstephenson/eco
② http://coffeekup.org/
③ http://www.mongodb.org/

我假设你已经安装好 MongoDB 了，如果你还没有，那么先安装[1]好再回来继续学习后续内容吧。

那么怎么让我们的应用使用 Mongo 呢？首先，需要安装 Mongoose[2]，一个流行的支持 Mongo 数据库的对象关系映射（object relational mapping，ORM）框架。直接通过 NPM 就可以很方便地安装 Mongoose：

```
> npm install mongoose

mongoose@2.4.7 ./node_modules/mongoose
--- hooks@0.1.9
--- colors@0.5.1
--- mongodb@0.9.7-2-1
```

> **提示**：从技术层面讲，Mongoose 并不算是个 ORM，因为 MongoDB 不是关系型数据库而是文档型数据库。不过，ORM 这个词如今所指的范围更广了，也可以指像 Mongoose 这类的工具。

安装好 Mongoose 之后，我们需要在应用中使用它。非常简单。首先创建一个 src/models/database.coffee 文件。src/models 目录用来存放所有和数据库相关的代码。

该文件内容如下。

例（源代码：app.3/src/models/database.coffee）

```
# Configure Mongoose (MongoDB):
global.mongoose = require('mongoose')
global.Schema = mongoose.Schema
global.ObjectId = Schema.ObjectId
mongoose.connect("mongodb://localhost:27017/csbook-todos")
```

上述代码中，我们首先引入了 Mongoose，然后，将该对象上的一些属性设置到了 global 对象上，为了后续在其他部分的代码中使用。

最后，我们配置了 Mongo 服务器的地址，以及数据库名。这部分代码可能要根据本地系统的具体配置情况进行修改。

最后要做的就是，在 app.coffee 中引入 src/models/database.coffee 文件，如下所示。

例（源代码：app.3/app.coffee）

```
# Setup Express.js:
global.express = require('express')
global.app = app = express.createServer()
require("#{__dirname}/src/configuration")

# Set up the Database:
require("#{__dirname}/src/models/database")
```

① http://www.mongodb.org/
② http://mongoosejs.com/

```
# Set up a routing for our homepage:
require("#{__dirname}/src/controllers/home_controller")

# Start server:
app.listen(3000)
console.log("Express server listening on port %d in %s mode", app.address().port,
➥app.settings.env)
```

本节的最后我们来创建一个 Todo 模型。该模型没有什么特别的，它应该有标题、ID、状态（用来表示是属于完成状态还是未完成状态），以及创建日期。

在 Mongoose 中我们使用 model 函数来创建新模型，传入该函数的是一个新的 Schema 对象。Schema 对象会告诉 Mongoose 以及 Mongo：模型中需要什么类型的数据。

例（源代码：app.3/src/models/todo.coffee）

```
# The Todo Mongoose model:
global.Todo = mongoose.model 'Todo', new Schema
  id: ObjectId
  title:
    type: String
    validate: /.+/
  state:
    type: String
    default: 'pending'
  created_at:
    type: Date
    default: Date.now
```

Todo 模型的 Schema 很简单。一看便知。对于 state 和 created_at 属性，我们定义了一些有用的默认值。对于 title 属性，我们设置了一个非常简单的校验，来确保它存储前符合不为空的要求。

> **提示：** Mongoose 还支持更为复杂的校验器，要了解更多这块信息，请参阅扩展文档[①]。

最后，我们需要更新 app.coffee 文件来引入刚刚创建的 Todo 模型。

例（源代码：app.3/app.coffee）

```
# Setup Express.js:
global.express = require('express')
global.app = app = express.createServer()
require("#{__dirname}/src/configuration")

# Set up the Database:
```

① http://mongoosejs.com/docs/validation.html

```
require("#{__dirname}/src/models/database")

# Set up a routing for our homepage:
require("#{__dirname}/src/controllers/home_controller")

# Start server:
app.listen(3000)
console.log("Express server listening on port %d in %s mode", app.address().port,
➥app.settings.env)
```

至此，数据库和数据模型都好了。在下一节中，你会看到，如何在我们所写的用来处理请求的控制器中操作数据模型。

10.3　编写待办事宜 API

最后剩下的服务器端的工作就是为下一章中要写的客户端定义 API，来访问待办事宜资源。我们新建一个文件来定义这些 API。

例（源代码：app.4/src/controllers/todos_controller.coffee）

```
# This 'controller' will handle the server requests
# for the Todo resource

# Get a list of the todos:
app.get '/api/todos', (req, res) ->
  res.json [{}]

# Create a new todo:
app.post '/api/todos', (req, res) ->
  res.json {}

# Get a specific todo:
app.get '/api/todos/:id', (req, res) ->
  res.json {}

# Update a specific todo:
app.put "/api/todos/:id", (req, res) ->
  res.json {}

# Destroy a specific todo:
app.delete '/api/todos/:id', (req, res) ->
  res.json {}
```

首先一点你应当注意到了，上述代码中所有的 API 我们都返回了一个空的 JSON[1]对象。稍后我会修改返回数据的代码，不过，首先来看看到目前为止我们已经完成了哪些。

① http://www.json.org/

本章最初部分，我提到过 Express 提供了 4 个函数，对应 4 种 HTTP 行为：get、post、put 和 delete。在 todos_controller.coffee 文件中，我们就使用了这些函数来让我们的 API 符合 REST[1]风格。

> **提示：** 我们可以只用 GET 和 POST 请求，但对我来说，这样会让我们的 API 变得令人困惑。而 REST 风格则会让我们定义的 API 更好，也更清晰。

我们需要对待办事宜执行 5 种操作。

第一种是从数据库中获取所有待办事宜列表。这就是第一个路由要定义的内容。

第二种是创建一个新的待办事宜。这里我们会使用 post 函数。

第三种是需要从数据库中获取指定的待办事宜。上述代码中，第三个路由就会映射到该操作。由于我们得要知道从数据库中获取哪个待办事宜，因此需要知道待办事宜的 ID。Express 允许写路由规则的时候使用占位符。然后我们可以在回调函数中获取占位符位置的具体信息。上述例子中我们用了:id。后续，通过使用 req 对象上的 param 函数就可以获取:id 的值。在更新和删除待办事宜时也会使用同样的方式。

第四种是更新指定的待办事宜。我们使用 put 函数，对应了 http 的 PUT 行为。

第五种是删除指定的待办事宜。

所有的处理结果都是返回一个 JSON 数据。Express 提供了一个 response 对象上的 json 方法，可以很方便地返回 JSON 数据。如果没有 json 方法的话，需要采用如下方式手动完成：

```
res.send JSON.stringify({})
```

接下来，我们来测试一下 API 的工作是否如我们所愿。启动应用之后，访问 http://localhost:3000/api/todos。你应当就能看到如下的信息显示在页面上：

```
[{}]
```

这意味着，我们列举所有待办事宜的 API 工作正常。

10.4 用 Mongoose 做查询操作

在开始应用的前端部分之前，本章最后要做的就是实现每一个 API 的内部操作。

10.4.1 查找所有待办事宜

让我们从数据库中获取所有的待办事宜。

---

① http://en.wikipedia.org/wiki/REST

例（源代码：app.5/src/controllers/todos_controller.coffee）

```
# This 'controller' will handle the server requests
# for the Todo resource

# Get a list of the todos:
app.get '/api/todos', (req, res) ->
  Todo.find {}, [], {sort: [["created_at", -1]]}, (err, @todos) =>
  if err?
    res.json(err, 500)
  else
    res.json @todos

# Create a new todo:
  app.post '/api/todos', (req, res) ->
    res.json {}

# Get a specific todo:
  app.get '/api/todos/:id', (req, res) ->
    res.json {}

# Update a specific todo:
  app.put "/api/todos/:id", (req, res) ->
    res.json {}

# Destroy a specific todo:
  app.delete '/api/todos/:id', (req, res) ->
    res.json {}
```

Mongoose 提供了许多有用的、方便的类级函数，用来帮助从数据库中查找指定的记录。在我们的例子中，用来查询数据库中所有的记录。

> **提示**：Mongoose 提供了非常好的文档。我强烈建议去 Mongoose 的网页上查阅一下关于查找文档[1]和创建复杂查询语句[2]的文档。

要查询所有的记录，可以简单调用 find 函数，传递一个回调函数给它即可。查询执行完毕后（不论成功还是失败）会调用回调函数。对于我们的例子而言，希望查询返回的结果能按时间顺序排好，将最近的放在最前面，最久的放在最后面。为此，我们必须传递一个包含这种排序细节的对象。这里有点巧妙。有两种实现方式：可以将查询语句切分为多行，首先构建一个对应的查询，然后再执行；或者采用现在这种方式，在包含排序细节的对象参数前，传递一个空对象作为查询参数。这里第一个参数是用来放置任何要查询的语句的。上述例子中，我们不想对查询结果做任何限制。第二个参数是一个数组，用来指定要获取的字段。由于我们需要所有字段，因此该参数也留空。

---

① http://mongoosejs.com/docs/finding-documents.html
② http://mongoosejs.com/docs/query.html

> **提示：**为什么在定义 find 调用的排序时要采用嵌套数组呢？事实上，这个是由 MongoDB 的性质决定的。如果想要根据多字段进行排序，可以采用类似{sort:[["created_at",-1], ["updated_at", 1]]}这样的方式。内部数组是一个组，表示了要添加的排序类型。其中，数组第一个元素是属性名，第二个是顺序，-1 表示降序，1 表示升序。其实在我看来它可以设计得更好更简洁，不过，目前它就是这样，没办法。

正如我此前提过的，传递给 find 函数的回调函数，不管查询结果成功与否都会被调用。这就意味着我们必须要自己来做错误处理。

回调函数接收两个参数。如果有错误的话，第一个参数是一个错误（Error）对象，否则就是 null。第二个参数是查询结果，上述例子中表示包含所有待办事宜对象的数组。如果查询失败，抛出错误，那么第二个参数会是 null。

在回调函数中，我们首先要检查是否有错误。可以使用 CoffeeScript 提供的有关存在的操作符——?，用来检查错误（err）对象是否不为 undefined 或者 null。如果有错误，就调用 json 函数，并传递错误信息和 500 状态码。这样可以让应用前端部分根据错误类型做相应的响应。

若没有错误，就可以将查询获得的待办事宜列表（@todos）传递给 json 函数，接下来的事情交给 Express 处理就好了。

> **提示：**你可能已经注意到了我们使用了=>来定义回调函数。这是为了确保能够访问到 req 和 res 对象。所有 Mongoose 的调用都是异步的，这就意味着，程序在执行回调函数前后会切换执行上下文，就有可能无法在回调函数中访问到 req 和 res 对象所在的作用域。

10.4.2 创建新的待办事宜

我们继续来看看如何创建新的待办事宜。

例（源代码：app.6/src/controllers/todos_controller.coffee）

```
# This 'controller' will handle the server requests
# for the Todo resource

# Get a list of the todos:
app.get '/api/todos', (req, res) ->
  Todo.find {}, [], {sort: [["created_at", -1]]}, (err, @todos) =>
    if err?
      res.json(err, 500)
    else
      res.json @todos

# Create a new todo:
app.post '/api/todos', (req, res) ->
  @todo = new Todo(req.param('todo'))
  @todo.save (err) =>
```

```
    if err?
      res.json(err, 500)
    else
      res.json @todo

# Get a specific todo:
app.get '/api/todos/:id', (req, res) ->
  res.json {}

# Update a specific todo:
  app.put "/api/todos/:id", (req, res) ->
    res.json {}

# Destroy a specific todo:
app.delete '/api/todos/:id', (req, res) ->
  res.json {}
```

创建新的待办事宜的代码非常简单。首先，我们初始化了一个新的待办事宜（Todo）类的实例，并将相关的数据对象传递了进去。这些数据哪里来的呢？我们从 POST 请求传递进来的参数中获取而来。

Express 允许通过 req 对象上的 param 函数来获取参数。我们将参数名传递给 param 函数，如果参数存在则返回参数值，否则就返回 null。

假设我们获取的待办事宜数据参数是一个键值对的对象，包含了创建新的待办事宜所需的数据，类似下面这样：

```
{
  todo: {
    title: "My Todo Title",
    state: "pending"
  }
}
```

要保存新创建的待办事宜，可以调用 Mongoose 提供的 save 函数，并传递一个回调函数。该回调函数就接收一个表示错误对象的参数，如果有错误的话。该回调函数和此前我们获取所有待办事宜时的类似。首先检查是否有错误，如果有，则响应错误内容以及 500 错误码，否则，就将新创建的待办事宜以 JSON 的形式发送回去。

> **提示**：从技术层面来讲，如果真的要支持 REST 协议，应当在创建了新的待办事宜后返回 201（“创建成功”）状态码。不过，Express 默认返回的状态码 200（“成功”）对我们来说足够了。实现 201 状态码就留给你自己去完成了。

10.4.3　获取、更新以及销毁待办事宜

我们的 API 中最后剩下的三种行为从本质上来说很相似，这里可以一起介绍。

例 （源代码：app.7/src/controllers/todos_controller.coffee）

```
# This 'controller' will handle the server requests
# for the Todo resource

# Get a list of the todos:
app.get '/api/todos', (req, res) ->
  Todo.find {}, [], {sort: [["created_at", -1]]}, (err, @todos) =>
    if err?
      res.json(err, 500)
    else
      res.json @todos

# Create a new todo:
app.post '/api/todos', (req, res) ->
  @todo = new Todo(req.param('todo'))
  @todo.save (err) =>
    if err?
      res.json(err, 500)
    else
      res.json @todo

# Get a specific todo:
app.get '/api/todos/:id', (req, res) ->
  Todo.findById req.param('id'), (err, @todo) =>
    if err?
      res.json(err, 500)
    else
      res.json @todo

# Update a specific todo:
app.put "/api/todos/:id", (req, res) ->
  Todo.findById req.param('id'), (err, @todo) =>
    if err?
      res.json(err, 500)
    else
      @todo.set(req.param('todo'))
      @todo.save (err) =>
        if err?
          res.json(err, 500)
        else
          res.json @todo

# Destroy a specific todo:
app.delete '/api/todos/:id', (req, res) ->
  Todo.findById req.param('id'), (err, @todo) =>
    if err?
      res.json(err, 500)
    else
      @todo.remove()
      res.json @todo
```

对于这三种行为，首先要做的都是根据:id 来查询到对应的待办事宜。这里我们使用

Mongoose 提供的类级函数 findById 来实现，并将 id 参数传递给它。

findById 函数接收一个回调函数，该回调函数一如既往地接收两个参数。如果存在错误的话，第一个参数就是错误对象，如果找到了要查询的待办事宜，第二个就是查询到的待办事宜。

对查询到的待办事宜如何处理要根据三种不同的行为而定。对于只需要用来展示的行为而言，我们只要以 JSON 格式返回该待办事宜数据即可。

对于要更新的行为，我们使用 set 函数，将要更新的属性传递给它。和我们创建新的待办事宜类似。

最后，对于销毁的行为，我们简单调用 remove 函数就能将数据从数据库中销毁了。

10.4.4　简化控制器

不知道你注意到没，我们的控制器充满了重复代码。就因为这只是个示例代码，我们就这样放任吗？为什么不做简化呢？借此还能再重温一下类的概念。

> **提示：**实际上，接下来要展示的简化的代码可能有点过分重构了，不过，我觉得能够展示一下如何能够根据意愿重构到这种程度，还是很有意思的。

要简化我们的控制器，首先要创建几个类对应这几种行为。还会用继承来减少重复代码，特别是在每种行为下都会遇到的错误处理这块。

我们从创建父类 Responder 开始，后面所有其他类都会继承此类。此类负责处理默认和通用的功能。

例（源代码：app.8/src/controllers/responders/responder.coffee）

```
class global.Responder

  respond: (@req, @res) =>
    @complete(null, {})

  complete: (err, result = {}) =>
    if err?
      @res.json(err, 500)
    else
      @res.json result
```

首先你应当注意到的是，我们将 Responder 类定义在了 global 对象上。这样可以在其他地方很方便地使用此类。

接下来，我们添加了一个默认的 respond 函数。这个函数会在 todos_controller.coffee 文件中对应的行为下被调用。respond 函数接收两个参数，req 对象和 res 对象。这里我们通过加上@前缀来自动将这些存储到类的作用域，这样我们可以在其他函数中访问这两个值。

尽管 respond 函数大部分情况下会被子类重写，为了以防万一我们还是提供了一个默认的行为——调用 complete 函数。

complete 函数包含了此前重复的错误处理，以及响应逻辑。

完成父类之后，我们开始写子类来处理对应的行为，从获取所有待办事宜的行为开始。

例（源代码：app.8/src/controllers/responders/index_responder.coffee）

```
require "#{__dirname}/responder"

class Responder.Index extends Responder

  respond: (@req, @res) =>
    Todo.find {}, [], {sort: [["created_at", -1]]}, @complete
```

首先，使用 Node 提供的 require 函数来引入此前实现的 Responder 类。

在定义类时，可以将类赋值给 global 对象，就像我们处理 Responder 类的方式一样。不过，这种方式并非最佳实践，应当多采用命名空间，而不是一股脑儿地将所有的东西都丢给 global 对象，所以接下来我们就要采用命名空间的方式。另外，我们要使用 CoffeeScript 的 extend 关键字来确保让 Responder.Index 类继承自 Responder。

> **提示**：在实际应用中，我们会给构建的类更好的命名，来清楚地指明它们是处理待办事宜资源的。我们目前这种方式，处理的资源一多就会有问题了。

最后，我们需要写一个自定义的 respond 方法。在该方法中，会包含此前实现的查询所有待办事宜的代码。与此前所写的相比较而言，最大的不同就是，传递的回调函数是在父类中定义的 complete 函数。

接着，我们来写一个处理创建新的待办事宜的类。

例（源代码：app.8/src/controllers/responders/create_responder.coffee）

```
require "#{__dirname}/responder"

class Responder.Create extends Responder

  respond: (@req, @res) =>
    todo = new Todo(@req.param('todo'))
    todo.save(@complete)
```

Responder.Create 类与我们刚刚构建的 Responder.Index 类非常相似。在 respond 函数中，创建新的待办事宜，并将 complete 函数作为调用 save 函数的回调函数传递进去。

例（源代码：app.8/src/controllers/responders/show_responder.coffee）

```
require "#{__dirname}/responder"

class Responder.Show extends Responder

  respond: (@req, @res) =>
    Todo.findById @req.param('id'), @complete
```

同样，Responder.Show 类和此前两个类也很相似，也是重写 respond 函数。我们会用 Responder.Show 类作为最后要创建的两个类的父类。让我们来看一下。

例（源代码：app.8/src/controllers/responders/update_responder.coffee）

```
require "#{__dirname}/show_responder"

class Responder.Update extends Responder.Show

  complete: (err, result = {}) =>
    if err?
      super
    else
      result.set(@req.param('todo'))
      result.save(super)
```

更新待办事宜的操作也需要先找到要更新的对象，所以我们这里继承 Responder.Show 类，而不是 Responder 类，这样做的好处就是，Responder.Show 类中有查找待办事宜所需的所有功能。

在 Responder.Update 类中，不需要写新的 respond 函数，直接从父类 Responder .Show 中继承即可。不过，我们需要自定义 complete 函数。

在 complete 函数中首先要做的就是检查是否有错误发生。如果有错误，就调用 super 函数，它会调用父类中的 complete 函数来进行错误处理。

如果没有错误，则设置要更新的字段并调用 save 函数。在调用 save 函数时，我们将 super 函数（也就是父类的 complete 函数）作为参数传递进去。它会对查询结果做相应处理。

销毁待办事宜的类和更新待办事宜的类非常相似。

例（源代码：app.8/src/controllers/responders/destroy_responder.coffee）

```
require "#{__dirname}/show_responder"

class Responder.Destroy extends Responder.Show

  complete: (err, result = {}) =>
    unless err?
      result.remove()
    super
```

同样，我们继承 Responder.Show 类，并且与 Responder.Update 类一样，需要自定义 complete 函数。

在 complete 函数中，首先需要检查在查询待办事宜时是否有错误。如果没有错误，则调用查询到的待办事宜对象上的 remove 函数，就可以将其从数据库中销毁了。

最后，我们直接调用 super 函数来处理错误和做出相应的响应。

剩下要做的就是，更新一下 todos_controller.coffee 文件来使用最新构建的类。

例 （源代码：app.8/src/controllers/todos_controller.coffee）

```
# require all of our responders:
for name in ["index", "create", "show", "update", "destroy"]
  require("#{__dirname}/responders/#{name}_responder")

# This 'controller' will handle the server requests
# for the Todo resource

# Get a list of the todos:
app.get '/api/todos', new Responder.Index().respond

# Create a new todo:
app.post '/api/todos', new Responder.Create().respond

# Get a specific todo:
app.get '/api/todos/:id', new Responder.Show().respond

# Update a specific todo:
app.put "/api/todos/:id", new Responder.Update().respond

# Destroy a specific todo:
app.delete '/api/todos/:id', new Responder.Destroy().respond
```

在上述代码最开始部分，我们引入了所有构建的类。

> **提示：**我们也可以为每个文件逐个写 require 语句，毕竟数量也不多，不过我更喜欢通过创建一个数组然后循环来构建 require 语句。这样更清楚，而且要再添加其他文件的话，只需在数组中添加即可，不需要复制粘贴 require 代码。

引入了所需的类后，我们直接创建对应类的实例，并将实例上的 respond 作为参数传递来取代原先的回调函数。

现在，todos_controller.coffee 文件更“干净”了，我们将所有的通用功能都抽离出去了。这就意味着，后期如果需要改变错误处理的方式，或者要添加其他的通用功能，只需修改一个文件即可。

10.5 小结

本章中，我们构建了待办事宜应用的服务器端部分。构建了一个 Express 应用，使用 Mongoose 增加了对 MongoDB 的支持，并且对代码进行了很好的重构。

下一章中，我们会在本章的基础上，构建一个“性感十足”[1]的前端。我们会采用 jQuery 来和服务器端通信。

①好吧，我承认我有点儿自嘲。认识我的人都知道，其实我根本就不擅长构建“性感十足”的前端。

第 11 章

示例：待办事宜列表第 2 部分（客户端，使用 jQuery）

在第 10 章中，我们已经完成了待办事宜应用的服务器端部分。本章中，我们会构建应用的客户端部分。过程中会采用诸如 Twitter[1]的 Bootstrap[2]以及 jQuery 这样有趣的技术。

11.1 用 Twitter 的 Bootstrap 来构建 HTML

我们从为应用构建基本的 HTML 和样式开始。这里，我们会采用 Twitter 的 Bootstrap。Bootstrap 其实就是简单的 CSS 和 JavaScript 文件集，能帮助快速构建应用。它提供了简单的栅格布局，能帮助对齐页面元素，还对表单、按钮、列表等基础控件做了漂亮的样式定义。我强烈推荐你去查看下 Bootstrap 项目，详细了解下它所提供的功能，因为我们这里只会用到它一小部分的功能。

我们首先需要更新 `src/views/index.ejs` 来使用 Bootstrap 以及自定义的 CSS。

例 （源代码：`app.1/src/views/index.ejs`）

```
<!DOCTYPE html>
<html>
  <head>
    <title>Todos</title>
    <link rel="stylesheet"
href="http://twitter.github.com/bootstrap/1.4.0/bootstrap.min.css">
    <%- css('/application') %>
  </head>
```

① http://twitter.com
② http://twitter.github.com/bootstrap/

```
  <body>

  </body>
</html>
```

如上述代码所示，我们外联了 Bootstrap 提供的 CSS 文件。另外，上述代码中还有一个比较特别的动态代码——css 函数，我们将/application 传递给它。这里其实是用来添加自定义 CSS 的，后面还会用来添加 CoffeeScript 文件。

我们会用到 Trevor Burnham 开发的 connect-assets[1]模块。该模块可以自动查找我们的 CSS 文件和 CoffeeScript 文件，并将其添加到 HTML 文件中。对于 CoffeeScript 文件，它会自动编译成 JavaScript，省去了我们手动去编译的步骤。

> **提示**：如果你熟悉 Ruby on Rails 中的 asset-pipeline，connect-assets 也是类似的，只是它用在 Express 应用中罢了。

要让 connect-assets 去查找静态资源，需要在应用的根目录下创建一个 assets 文件夹，将静态资源都放在这里。与此同时，我们在其中写一点 CSS 来给 HTML 定义基础样式。

例（源代码: app.1/assets/application.css）

```
#todos li {
  margin-bottom: 20px;
}

#todos li .todo_title {
  width: 800px;
}

#todos li .completed {
  text-decoration: line-through;
}

#todos #new_todo .todo_title{
  width: 758px;
}
```

本书不是一本介绍 CSS 的书，所以我不会对上述代码做过多解释。如果你不理解上述代码，可以先做个标记，先来看看效果是怎样的。

现在我们需要安装 connect-assets 模块，以便应用刚写的代码：

```
> npm install connect-assets

connect-assets@2.1.6 ./node_modules/connect-assets
__ connect-file-cache@0.2.4
__ underscore@1.1.7
__ mime@1.2.2
__ snockets@1.3.3
```

① https://github.com/trevorBurnham/connect-assets

最后，我们需要告诉 Express 使用 connect-assets 模块来负责静态资源。怎么做呢？只需在 configuration.coffee 文件中添加一行代码即可。

例（源代码：app.1/src/configuration.coffee）

```
# Configure Express.js:
app.configure ->
  app.use(express.bodyParser())
  app.use(express.methodOverride())
  app.use(express.cookieParser())
  app.use(express.session(secret: 'd19e19fd62f62a216ecf7d7b1de434ad'))
  app.use(app.router)
  app.use(express.static(__dirname + '../public'))
  app.use(express.errorHandler(dumpExceptions: true, showStack: true))
  app.set('views', "#{__dirname}/views")
  app.set('view engine', 'ejs')
  app.use(require('connect-assets')())
```

如果现在启动应用的话，在访问 http://localhost:3000 时，会看到一张空白页，事实上也应该就是一张空白页。如果看到了其他内容，那肯定有地方出问题了。

在本节的最后，我们来添加一个表单，用来创建新的待办事宜。这非常有用，因为目前数据库中没有任何数据，一切都将从这里开始。

例（源代码：app.2/src/views/index.ejs）

```
<!DOCTYPE html>
<html>
  <head>
    <title>Todos</title>
      <link rel="stylesheet" href="http://twitter.github.com/bootstrap/1.4.0/bootstrap.
➥min.css">
      <%- css('/application') %>
  </head>
  <body>
    <div class="container">
      <h1>Todo List</h1>
      <ul id='todos' class='unstyled'>
        <li id='new_todo'>
          <div class="clearfix">
            <div class="input">
              <div class="input-prepend">
                <span class='add-on'>New Todo</span>
                <input class="xlarge todo_title" size="50" type="text"
➥placeholder="Enter your new Todo here..." />
            </div>
          </div>
        </div>
      </li>
    </ul>
```

```
</div>

</body>
</html>
```

现在再启动应用访问的话，就能看到一个很漂亮的表单，可以输入新待办事宜的数据。不过，目前表单没有任何效果，后续会改进。你可能想知道这些样式都是哪里来的——它们都来自于此前我们介绍过的 Bootstrap。

11.2　使用 jQuery 进行交互

现在已经有了表单了，接下来给它添加对应的功能。这里我们使用 jQuery，一个非常优秀的、被广泛使用的库。对我来说，在 JavaScript 的生态圈中，没有比 jQuery 更强大的工具集了。jQuery 最初是 John Resig[①]在 2006 年发布的，目前为止，全球前 10000[②]家网站中有 49%的网站都在使用它。它也是当前最流行的 JavaScript 库之一，jQuery 可以让我们在以下方面写出非常精简的代码：操作 HTML DOM、执行 AJAX 请求、处理事件以及简单动画。它还是跨平台的，这意味着它能在所有主流的操作系统和浏览器上工作。

> **提示**：在 Web 混沌黑暗的时代，开发者不得不为同样一段 JavaScript 编写多个版本。一个版本工作在 IE 浏览器中、一个版本工作在网景浏览器中，等等。而现如今，有了 jQuery，你只需写一份代码即可，jQuery 会确保它能够在所有主流浏览器中都能工作。

将 jQuery 用到我们的应用中非常简单，只需要在 `index.ejs` 中引用即可。

例（源代码：`app.3/src/views/index.ejs`）

```
<!DOCTYPE html>
<html>
  <head>
    <title>Todos</title>
    <script src="http://ajax.googleapis.com/ajax/libs/jquery/1.7.1/jquery.min.js"
➥type="text/javascript"></script>
    <%- js('/application') %>

    <link rel="stylesheet" href="http://twitter.github.com/bootstrap/1.4.0/
➥bootstrap.min.css">
    <%- css('/application') %>
  </head>
  <body>

    <div class="container">
      <h1>Todo List</h1>
      <ul id='todos' class='unstyled'>
        <li id='new_todo'>
          <div class="clearfix">
```

① http://en.wikipedia.org/wiki/John_Resig
② http://en.wikipedia.org/wiki/Jquery

```
          <div class="input">
            <div class="input-prepend">
              <span class='add-on'>New Todo</span>
              <input class="xlarge todo_title" size="50" type="text"
➥placeholder="Enter your new Todo here..." />
            </div>
          </div>
        </div>
      </li>
    </ul>
  </div>

</body>
</html>
```

提示：在实际应用中，不鼓励像上述代码这样引入外部的库。因为那些库很有可能会更新，会有 bug 影响到应用本身；甚至更糟的是，外部库的引用可能会被移除或者不可用，这样一来应用也会挂掉。推荐将库拷贝到本地。不过，对于本书来说，这样直接引用简单很多。

11.3　给新建待办事宜表单添加功能

除了添加 jQuery 到我们的应用中之外，还用了 connect-assets 提供的 js 函数来引入 application 文件，这里 application 文件指的就是 assets/application.coffee。

例（源代码：app.3/assets/application.coffee）

```
#= require_tree "jquery"
```

application.coffee 文件非常简短，目前先保持这样。这里用到了 connect-assets 允许引入其他 CoffeeScript 文件或 JavaScript 文件的特性。在上述例子中，我们引入了一个名为 jquery 的目录。这意味着，所有在该目录下创建的文件都会被自动引入进来。

在 assets 文件夹中，创建一个名为 jquery 的文件夹，并在该文件夹中创建一个 new_todo_form.coffee 文件。我们会将 HTML 中所有处理创建新待办事宜表单的代码都放在该文件中。下面我们就来看看这部分代码。

在开始写代码前，先来看看我们的需求是怎么样的。当用户填写完新的待办事宜信息并按下回车键时，先要检测数据是否有效，这里要求至少有一个非空白字符。如果无效，就指出错误让用户知道问题出在哪里。如果有效，就通过 POST 请求将数据发送给对应的 API。如果来自 API 的响应结果是成功，就将新的待办事宜添加到待办事宜列表中并重置表单。如果响应结果是失败，就将错误消息显示给用户。下面是具体的代码。

例（源代码：app.3/assets/jquery/new_todo_form.coffee）

```
$ ->
  # Focus the new todo form when the page loads:
  $('#new_todo .todo_title').focus()
```

```
# Handle the keypress in the new todo form:
$('#new_todo .todo_title').keypress (e) ->
  # we're only interested in the 'enter' key:
  if e.keyCode is 13
    todo = {title: $(e.target).val()}
    if !todo.title? or todo.title.trim() is ""
      alert "Title can't be blank"
    else
      request = $.post "/api/todos", todo: todo
      request.fail (response) =>
        message = JSON.parse(response.responseText).message
        alert message

      request.done (todo) =>
        $('#new_todo').after("<li>#{JSON.stringify(todo)}</li>")
        $(e.target).val("")
```

页面载入完毕后要做的第一件事就是将焦点聚焦到表单上。这样做可以让用户很方便地直接就开始输入新的待办事宜信息，不需要手动先去聚焦。

> **提示：** 在 jQuery 中，调用`$`函数，并传递一个回调函数，该回调函数会在整个页面载入完毕后才会被调用到。而且，可以以这种方式调用多次`$`函数，很方便。

接下来，我们给表单添加函数，每当用户按下键盘上的键时，就会触发该处理函数。这里我们只需要处理回车键，回车键的键码是 13，所以先检查 `event` 对象的 `keyCode` 属性是否为 13，如果是就紧接着做相应处理，否则直接忽略该事件。

如果键码为 13，就可以进一步获取表单的值——我们要新建的待办事宜的标题。然后，创建一个包含服务器端所需数据的对象。

获取到标题后，我们对其做一些本地校验。这样做的好处在于本地校验比服务器端校验速度更快，有更好的用户体验。服务器端校验要先发送数据给服务器端、等待校验结果、再返回给客户端，并且这过程中还有可能出错。

> **提示：** 有许多方式可以做到一套校验两端使用（客户端和服务器端），这样可以不需要写两套校验代码。如果你也能找到这样的方式，当然再好不过。不过，并非所有的校验工作两端都可以做。最明显的例子就是用户名的校验，必须要走服务器端来检查用户名是否唯一。可以使用 AJAX 和 API 调用来校验；也可以在前端校验用户名是否为空，然后再交给服务器端做进一步校验。

假设本地校验通过，就可以将数据返回给 API 了。我们使用 jQuery 提供的 `Post` 函数创建一个新的 AJAX 请求。同时提供了请求 `url`（`/api/todos`）以及一个包含要发送数据的对象。我们把 `Post` 函数的返回值赋给了一个名为 `request` 的变量，后面会将请求对应的回调函数绑定到该变量上。

> **提示：** jQuery 1.5 中引入了 `deferred` 对象。在这之前，必须要将回调函数作为参数传递给要调用的函数，如 `Post` 或者 `ajax`。非常受限。有了 `deferred` 对象，可以在任何时候绑定回调函数，甚至可以在请求完成后再绑定回调函数。这让写出更加独立的代码变得更加容易。

上述例子中，我们在 request 对象上添加了两个回调函数。第一个回调函数用来处理请求失败的情况。该回调函数接收一个 response 对象作为参数，通过该对象，我们要进一步获取对象上的 responseText 属性，该属性值是 JSON 字符串，解析该字符串之后，获取其中的错误信息，并将其显示出来。

第二个回调函数用来处理请求成功（成功在数据库中创建了新的待办事宜）的情况。在函数内部，我们没有写一个漂亮的模板来显示新创建的待办事宜，取而代之的是，直接将数据以 JSON 字符串的形式添加在 li 标签中，然后追加到待办事宜列表中，并显示在表单后面，这样就能确保最新的待办事宜永远在最上面。

如果你将应用运行起来，尝试创建一个新的待办事宜，就应当能够看到类似下面这样的信息展现在表单的下面：

```
{"title":"My New Todo","_id":"4efa82bdf65049000000001a","created_at":"2011-12-
➥28T02:45:17.992Z","state":"pending"}
```

使用 underscore.js 模板来重构待办事宜列表

有许多 JavaScript 的模板引擎，为自己的应用选择合适的模板引擎至关重要。由于我们的需求很简单，所以我会选择一款简单的模板引擎。我们将使用 underscore.js[①]。为什么在这么多模板引擎中偏偏选择它呢？原因很简单，underscore.js 是我们后面要用到的 Backbone.js 库所依赖的，所以反正都是要引入的，还不如现在就用呢。

> **提示：**我个人喜欢一个叫 eco[②]的模板引擎，它允许在模板中直接嵌入 CoffeeScript 脚本。其他流行的模板引擎还有 Handlebars[③]、Mustache[④]以及 Jade[⑤]。jQuery 的模板插件[⑥]也是一个很不错的选择，可惜它已经废弃了，不再维护和开发了，因此，如果你打算用它，我建议你打消这个念头，尽早选择一款其他能满足需求的模板引擎。

我们来更新下 index.ejs 文件，将 underscore.js 引入进来。

例（源代码：app.4/src/views/index.ejs）

```
<!DOCTYPE html>
<html>
  <head>
    <title>Todos</title>
    <script src="http://ajax.googleapis.com/ajax/libs/jquery/1.7.1/jquery.min.js"
➥type="text/javascript"></script>
    <script src="http://documentcloud.github.com/underscore/underscore-min.js"
```

① http://documentcloud.github.com/underscore
② https://github.com/sstephenson/eco
③ http://www.handlebarsjs.com/
④ https://github.com/janl/mustache.js/
⑤ http://jade-lang.com/
⑥ https://github.com/jquery/jquery-tmpl

```
➥type="text/javascript"></script>
    <%- js('/application') %>

    <link rel="stylesheet" href="http://twitter.github.com/bootstrap/1.4.0/
➥bootstrap.min.css">
    <%- css('/application') %>
  </head>
  <body>

  <div class="container">
    <h1>Todo List</h1>
    <ul id='todos' class='unstyled'>
      <li id='new_todo'>
        <div class="clearfix">
          <div class="input">
            <div class="input-prepend">
              <span class='add-on'>New Todo</span>
              <input class="xlarge todo_title" size="50" type="text"
➥placeholder="Enter your new Todo here..." />
            </div>
          </div>
        </div>
      </li>
    </ul>
  </div>

  </body>
</html>
```

这里还少做了一件事情，我们稍后会在介绍 Backbone 的时候做说明。

接下来，我们创建一个新的 assets/templates.coffee 文件，该文件中包含了应用所需的模板。

例（源代码：app.4/assets/templates.coffee）

```
# Change the syntax for underscore.js templates.
# The pattern is now {{some_var}} instead of <%= some_var %>
_.templateSettings =
  interpolate : /\{\{(.+?)\}\}/g

@Templates = {}

Templates.list_item_template = """
<div class="clearfix">
  <div class="input">
    <div class="input-prepend">
      <label class="add-on active"><input type="checkbox"
➥class="todo_state" /></label>
      <input class="xlarge todo_title" size="50" type="text" value="{{title}}" />
      <button class='btn danger'>X</button>
    </div>
  </div>
</div>
"""
```

上述代码中，我们首先更改了 underscore.js 中默认的动态数据占位符。原因是我发现{{ }}输入起来比<%= %>更容易。同时，这也是绝大多数模板引擎所使用的，这样以后要是换个模板引擎也很简单，不需要再去对模板做更改了。

上述代码最后是定义模板。我们创建了一个新的 Templates 对象，并为其添加了一个名为 list_item_template 的保存了 HTML 模板的属性，这里，创建模板使用了第 2 章中介绍的 CoffeeScript 中的 heredoc 语法。

模板非常简单，要动态插入的数据就只有待办事宜的标题。外部的勾选框用来更新状态。下面我们在 assets/jquery 文件夹下创建一个 todo_item.coffee 文件。在该文件中创建一个函数，使用刚刚创建的模板来将一个新的待办事宜添加到列表中。

例（源代码：app.4/assets/jquery/todo_item.coffee）

```
@TodoApp ||= {}

TodoApp.appendTodo = (todo) ->
  li = $("<li>#{_.template(Templates.list_item_template)(todo)}</li>")
  $('#new_todo').after(li)
```

首先，我们创建了一个可供外部访问的新变量 TodoApp。通过在变量名前加上@缀，就能将该变量绑定到 this 对象上，在上述例子中，this 就是浏览器中的 window 对象。然后，该变量就可以在其他 window 作用域下被访问到了。

接下来，我们创建了一个函数，接收一个待办事宜，并将其用模板显示到屏幕上。underscore 库提供一个_变量，通过该变量可以调用 template 函数。调用该函数时，我们将定义好的模板以及 JSON 格式的数据传递给它。underscore 库会完成剩下的工作，并返回我们想要的 HTML。之后，我们将返回的 HTML 片段添加到待办事宜列表中，并显示在表单后面。

现在，我们更新 new_todo_form.coffee 文件，在创建新待办事宜时，使用刚刚新创建的函数。

例（源代码：app.4/assets/jquery/new_todo_form.coffee）

```
$ ->
  # Focus the new todo form when the page loads:
  $('#new_todo .todo_title').focus()

  # Handle the keypress in the new todo form:
  $('#new_todo .todo_title').keypress (e) ->
    # we're only interested in the 'enter' key:
    if e.keyCode is 13
      todo = {title: $(e.target).val()}
      if !todo.title? or todo.title.trim() is ""
        alert "Title can't be blank"
      else
        request = $.post "/api/todos", todo: todo
        request.fail (response) =>
```

```
      message = JSON.parse(response.responseText).message
      alert message
    request.done (todo) =>
      TodoApp.appendTodo(todo)
      $(e.target).val("")
```

在 request.done 回调函数中，我们将直接以 JSON 字符串形式打印出从服务器端获取到的待办事宜的代码，替换为调用之前我们新创建的 TodoApp.appendTodo 函数。

重启应用，添加一个新的待办事宜，这次你应当能够看到在待办事宜列表中添加了一个具有良好样式的新待办事宜。

11.4 列举现有的待办事宜

至此，我们已经可以在数据库中创建新的待办事宜，并且可以在创建完之后将其显示到待办事宜列表中。下面需要实现在载入页面时，将现有的待办事宜展示出来。因为目前我们在应用中创建好的待办事宜，在刷新页面之后，好像就不见了。实际上，它们都在数据库中，只是我们还没有实现将其从数据库中获取出来并显示到页面上。幸运的是，实现该功能非常容易。

例（源代码：app.5/assets/jquery/retrieve_existing_todos.coffee）

```
$ ->
  request = $.get('/api/todos')
  request.done (todos) ->
    for todo in todos.reverse()
      TodoApp.appendTodo(todo)
```

首先，我们等到页面完全载入完毕之后，从 API 获取待办事宜。然后，创建一个 request 对象，指向获取待办事宜的 API，并绑定一个回调函数，来处理成功获取到的待办事宜。在回调函数中，我们将待办事宜倒序排列之后，逐个传递给 appendTodo 函数，这里为什么要倒序排列呢？原因很简单，不过也很无奈。事实上从服务器端返回的待办事宜的顺序是正确的，只是因为我们的 appendTodo 函数是将每个待办事宜添加到列表的顶端，导致顺序反过来了，所以，我们要先进行倒序排列下。当然，我们也可以改变 API 工作的方式，不过这样做显然不对。因为是客户端的操作导致顺序反了，不应该让服务器端来解决。我们也可以再写一个函数来将它们正确排序，或者在 appendTodo 函数中根据具体条件进行排序，不过，不管怎么说，目前这种解决方式最简单，也最不容易出问题。

现在再重新运行我们的应用，就应该能看到所有创建在数据库里的现有的待办事宜了。

11.5 更新待办事宜

创建新的待办事宜并且将现有的待办事宜都展示出来都可以了，接下来，我们要实现对待办事宜的修改。在我们的应用中，需要更新待办事宜的 Title 属性以及改变其状态。

我们来添加两个函数。第一个用来监听每个待办事宜中复选框以及输入框的变化。有任何变化发生，就会调用第二个函数——调用服务器端更新待办事宜的 API。

例（源代码：app.6/assets/jquery/watch_todo_for_changes.coffee）

```
@TodoApp ||= {}

# Watch the todo for changes:
TodoApp.watchForChanges = (li, todo) ->
  # If the checkbox is checked/unchecked:
  $('.todo_state', li).click (e) =>
    TodoApp.updateTodo(li, todo)
  # If someone hits "enter" in the title field:
  $('.todo_title', li).keypress (e) =>
    if e.keyCode is 13
      TodoApp.updateTodo(li, todo)
```

顾名思义，watchForChanges 会去监听待办事宜对应的 li 标签，如果复选框被勾选或是取消，或者用户在标题输入框中按下了回车键，就会触发 updateTodo 函数。为了确保 watchForChanges 函数能被调用到，我们在 appendTodo 函数中追加一个新的待办事宜时，要更新 watchForChanges 函数。

例（源代码：app.6/assets/jquery/todo_item.coffee）

```
@TodoApp ||= {}

TodoApp.appendTodo = (todo) ->
  li = $("<li>#{_.template(Templates.list_item_template)(todo)}</li>")
  $('#new_todo').after(li)
  TodoApp.watchForChanges(li, todo)
```

监听待办事宜之后，我们来写一个 updateTodo 函数将这些变化保存下来。

例（源代码：app.6/assets/jquery/update_todo.coffee）

```
@TodoApp ||= {}

# Update the todo:
TodoApp.updateTodo = (li, todo) ->
  todo.title = $('.todo_title', li).val()
  if !todo.title? or todo.title.trim() is ""
    alert "Title can't be blank"
  else
    if $('.todo_state', li).attr('checked')?
      todo.state = 'completed'
    else
      todo.state = 'pending'
    request = $.post "/api/todos/#{todo._id}",
      todo: todo
```

```
      _method: 'put'
    request.fail (response) =>
      message = JSON.parse(response.responseText).message
      alert message
```

updateTodo 函数和此前创建新待办事宜的函数类似。但是，也有几个重要的区别。首先，我们从标题文本输入框中获取值，并做简单的校验，确保其值不为空。如果校验通过，继续准备其他要发送给服务器端的数据。上述例子中，我们需要根据复选框是否选中来相应地更新待办事宜 state 属性的值。

接下来，我们向 API 发起了更新待办事宜的请求。由于是更新指定的待办事宜，我们必须确保在 API 的 URL 中包括该待办事宜的 ID。你可能已经注意到了，我们使用了 POST 来向服务器发送数据，而 API 指定的是要用 PUT。之所以这么做有几个原因。首先是历史原因，并非所有的浏览器都支持除 GET 和 POST 以外的其他 HTTP 行为；jQuery 也是如此。为了绕过这个限制，许多像 Express 以及 Ruby on Rails 这样的框架，都提供了一个专门的_method 参数。框架会根据该参数来判断请求类型。上述例子中，我们指定的_method 为 PUT，因此，Express 会将此请求视为 PUT 请求而不是 POST 请求。

最后，我们检查更新待办事宜时服务器端有没有发生错误。如果服务器端返回有错误，则将错误显示出来；否则，什么都不做，让用户继续处理他们的待办事宜。

不过，这里还有个小问题。如果我们将某个待办事宜标记为完成（completed）并刷新页面，你会发现它看起来好像还在处理状态中，原因是我们没有处理对应的复选框，而且也没有用相应的 completed 样式来标明让用户看到该待办事宜已经完成了。下面，我们就来解决这个问题，写一个函数来酌情更新这些样式。

例（源代码：app.7/assets/jquery/style_by_state.coffee）

```
@TodoApp ||= {}
# Update the style based on the state:
TodoApp.styleByState = (li, todo) ->
  if todo.state is "completed"
    $('.todo_state', li).attr('checked', true)
    $('label.active', li).removeClass('active')
    $('.todo_title', li).addClass('completed').attr('disabled', true)
  else
    $('.todo_state', li).attr('checked', false)
    $('label', li).addClass('active')
    $('.todo_title', li).removeClass('completed').attr('disabled', false)
```

上述函数非常简单。调用它时，它会检查待办事宜的状态，如果状态被标记为 completed，就会应用对应的样式。否则，将样式移除。这里，有一个 jQuery 的亮点，就是它不需要我们去检查元素上是否已经有了要添加的样式。如果已经有了，jQuery 会自动忽略添加请求。对于移除样式也是同样的，如果要移除的样式根本就没有，就什么都不会做。

最后，我们需要在相应的地方调用 styleByState 函数。首先，需要在 appendTodo 函数中调用该函数。在新的待办事宜添加到待办事宜列表时，要确保有正确的样式。

例（源代码：app.7/assets/jquery/todo_item.coffee）

```
@TodoApp ||= {}

TodoApp.appendTodo = (todo) ->
  li = $("<li>#{_.template(Templates.list_item_template)(todo)}</li>")
  $('#new_todo').after(li)
  TodoApp.watchForChanges(li, todo)
  TodoApp.styleByState(li, todo)
```

另外一个要调用该函数的地方是更新待办事宜的时候。我们可以通过在 updateTodo 函数中给 request 对象绑定一个回调函数来实现：

例（源代码：app.7/assets/jquery/update_todo.coffee）

```
@TodoApp ||= {}

# Update the todo:
TodoApp.updateTodo = (li, todo) ->
  todo.title = $('.todo_title', li).val()
  if !todo.title? or todo.title.trim() is ""
    alert "Title can't be blank"
  else
    if $('.todo_state', li).attr('checked')?
      todo.state = 'completed'
    else
      todo.state = 'pending'
    request = $.post "/api/todos/#{todo._id}",
      todo: todo
      _method: 'put'
    request.fail (response) =>
      message = JSON.parse(response.responseText).message
      alert message
    request.done (todo) ->
      TodoApp.styleByState(li, todo)
```

11.6 删除待办事宜

我们的应用快要完成了，最后就剩下添加 delete 按钮让用户能删除不想要的待办事宜的功能了。尽管实现该功能代码非常简单，但我们还是来看下吧。

我们从 deleteTodo 函数开始。

例（源代码：final/assets/jquery/delete_todo.coffee）

```
@TodoApp ||= {}

# Delete the todo:
TodoApp.deleteTodo = (li, todo) ->
  if confirm "Are you sure?"
    request = $.post "/api/todos/#{todo._id}", _method: 'delete'
```

```
    request.done =>
      li.remove()
```

deleteTodo 函数通过设置此前介绍的_method 参数，向 API 发起删除待办事宜的 DELETE 请求。如果删除成功，就将其从页面中删除——非常简单、清楚。

接下来，需要给 delete 按钮添加 click 事件来调用 deleteTodo 函数。为此，我们来修改此前写的 watchForChanges 函数。

例（源代码：final/assets/jquery/watch_todo_for_changes.coffee）

```
@TodoApp ||= {}

# Watch the todo for changes:
TodoApp.watchForChanges = (li, todo) ->
  # If the checkbox is checked/unchecked:
  $('.todo_state', li).click (e) =>
    TodoApp.updateTodo(li, todo)
  # If someone hits "enter" in the title field:
  $('.todo_title', li).keypress (e) =>
    if e.keyCode is 13
      TodoApp.updateTodo(li, todo)
  $('button.danger', li).click (e) =>
    e.preventDefault()
    TodoApp.deleteTodo(li, todo)
```

好了！完工！恭喜恭喜！

11.7　小结

本章，我们使用 jQuery 为待办事宜列表应用写了一个可交互的 Web 客户端。实现非常简单，并且你能看到 CoffeeScript 是如何帮助我们写出非常漂亮的 jQuery 代码的。

本章，我们从代码“体系结构”角度出发，采用了和 jQuery 工程师一样的方式。写了一堆函数，传递了很多对象，完成了必要的功能。我们也可以采取另外一种方式来实现：编写几个类来管理每个待办事宜，更简洁地将 HTML 元素封装起来并为其注册相应的事件。

那么，为什么本章不展示第二种实现方式呢？原因有二。其一，我已经提过，本章采取的方式是写原生 JavaScript 和 jQuery 代码常见的方式，因此，我想让你感受下这种方式用 CoffeeScript 写是怎么样的。其二，如果再以“类”的风格写代码，就等于在重复造轮子了，因为 Backbone.js 就提供了这种方式。

Backbone.js 是一个简单的框架，它允许将视图和页面元素对应起来，并将其和模型建立连接，比如我们的待办事宜，并且可以很容易地互相监听和响应事件。我们在第 12 章就会带你实践 Backbone。

顺便说下，如果你想看看如果我们用类的方式来实现本章介绍的客户端会是怎么样的，我已经写好了[1]，供你参阅！

---

① https://github.com/markbates/Programming-In-CoffeeScript/tree/master/todo2/alt-final

第 12 章

示例：待办事宜列表第 3 部分（客户端，使用 Backbone.js）

在第 11 章中，我们使用 jQuery 库构建了应用的客户端部分，并在第 11 章最后提过如何采用类的形式来实现，不过，我觉得这样做基本上是在重写 Backbone.js[1]框架。我是典型的讨厌重复造轮子的人，因此，我想我们应该看看用 Backbone 代替 jQuery 来实现应用的客户端会是怎么样的。

12.1 什么是 Backbone.js

Backbone 是用 JavaScript 开发的客户端 MVC[2]框架，其作者是 Jeremy Ashkenas[3]，他同时也是 CoffeeScript[4]的作者。Backbone 可以让我们用 JavaScript 或者 CoffeeScript 写出高响应的客户端应用。

Backbone 由三部分组成。第一部分是视图层。视图层负责页面元素的渲染，并监听元素的变化来做出相应的响应。还可以监听来自其他对象的事件，并根据事件详情更新对应的元素。

第二部分是模型和集合。一个模型对应一个实例对象。对于我们的例子来说，就是一个待办事宜。模型对象可以和数据存储中心进行通信并将自己持久化起来。它还包括其他一些方便操作对象的函数，比方说：将人名的姓和名拼接成一个字符串的函数。集合，顾名思义，是模型对象的集合，对于我们的例子来说就是 Todo 模型的集合。在 Backbone 中，集合也可以和数据存储中心进行通信，比方说我们的 API。

在 Backbone 中，当对象发生不同行为时，模型和集合还可以广播各种事件。比方说，当新的模型添加到集合中时，集合就会触发 add 事件。这些事件也可以被其他对象监听，比如视

① http://documentcloud.github.com/backbone/
② http://en.wikipedia.org/wiki/Model–view–controller
③ https://github.com/jashkenas/
④ 说句心里话，我在本章中特别介绍 Backbone 并非因为 Jeremy 是 CoffeeScript 的作者，而是因为 Backbone 真的很好，而且我一直在用。

图对象。拿视图对象来说，它会监听集合的 add 事件，当事件触发时，它会将模型以适当的方式渲染到界面上。后面会展示具体如何实现。

最后一部分是路由器。Backbone 路由器允许我们监听和响应浏览器 URL 的变化。当 URL 发生变化时，在路由器中配置的相关代码就会执行。这和我们此前用 Express 构建 API 时类似。在本章中，我们不会用到路由器，但这不代表路由器没什么用。只是我们的应用不需要而已。

至此，我们大概了解了什么是 Backbone。接下来，我们会更加详细地介绍 Backbone，不过本章不是全面介绍 Backbone 的教程，我们只会对应用所需的部分做相关介绍。

简化客户端代码

在重构应用之前，我们需要先简化下原有应用的代码，以便更好地将 Backbone 融入进去。首先要简化的是删除 `assets/jquery` 目录。

接下来，我们需要从 `assets/application.coffee` 文件中移除对该目录的引用。

例（源代码：`app.1/assets/application.coffee`）

```
#= require "templates"
```

好了！现在如果重启应用的话，就只能看到新建待办事宜的表单了，并且它完全是不起作用的。下面就让我们来重构我们的应用吧。

12.2　配置 Backbone.js

在应用中安装 Backbone 比较简单，它唯一依赖的就是：underscore.js[1]。尽管看起来 Backbone 只有这么一个“强”依赖，但是事实上，如果没有操作 DOM 和 AJAX 的库协助的话，Backbone 是发挥不了威力的，所以还需要 jQuery（或者 Zepto[2]）。好在我们在 `index.ejs` 中已经引入了 jQuery 和 underscore.js，所以我们只需添加 Backbone 本身就基本完成了。

例（源代码：`app.1/src/views/index.ejs`）

```
<!DOCTYPE html>
<html>
  <head>
    <title>Todos</title>
    <script src="http://ajax.googleapis.com/ajax/libs/jquery/1.7.1/jquery.min.js"
➥type="text/javascript"></script>
    <script src="http://documentcloud.github.com/underscore/underscore-min.js"
➥type="text/javascript"></script>
    <script src="http://documentcloud.github.com/backbone/backbone-min.js"
➥type="text/javascript"></script>
```

① http://documentcloud.github.com/underscore

② http://zeptojs.com

```
    <%- js('/application') %>

    <link rel="stylesheet"
➥href="http://twitter.github.com/bootstrap/1.4.0/bootstrap.min.css">
    <%- css('/application') %>
  </head>
  <body>

    <div class="container">
      <h1>Todo List</h1>
      <ul id='todos' class='unstyled'>
        <li id='new_todo'>
          <div class="clearfix">
            <div class="input">
              <div class="input-prepend">
                <span class='add-on'>New Todo</span>
                <input class="xlarge todo_title" size="50" type="text" placeholder="Enter
➥your new Todo here..." />
              </div>
            </div>
          </div>
        </li>
      </ul>
</div>

</body>
</html>
```

这就是在应用中配置运行 Backbone 要做的全部工作。不过，我还需要在启动文件中再多加一个文件。在写 Todo 模型之前，我不想加这个文件，不过这看起来像是一个把所有准备工作都就绪的好时机。

当 Backbone 和我们的 API 通信时，默认会发送类似如下格式的数据：

```
{title: 'My New Todo'}
```

不过，如果你还记得的话，我们的 API 希望数据是在某个命名空间下的，就像这样：

```
todo: {title: 'My New Todo'}
```

怎么办呢？我们需要从 Ruby Gem 中“借用”一个文件——backbone-rails[①]，它会给 Backbone 打个补丁来满足我们的需求。下面是该文件的内容。

例　（源代码：app.2/assets/backbone_sync.js）

```
// Taken from https://github.com/codebrew/backbone-rails.
// This namespaces the JSON sent back to the server under the model name.
// IE: {todo: {title: 'Foo'}}
(function() {
```

① https://github.com/codebrew/backbone-rails

```
var methodMap = {
  'create': 'POST',
  'update': 'PUT',
  'delete': 'DELETE',
  'read' : 'GET'
};

var getUrl = function(object) {
  if (!(object && object.url)) return null;
  return _.isFunction(object.url) ? object.url() : object.url;
};

var urlError = function() {
  throw new Error("A 'url' property or function must be specified");
};

Backbone.sync = function(method, model, options) {
  var type = methodMap[method];

  // Default JSON-request options.
  var params = _.extend({
    type:       type,
    dataType:   'json',
    beforeSend: function( xhr ) {
      var token = $('meta[name="csrf-token"]').attr('content');
      if (token) xhr.setRequestHeader('X-CSRF-Token', token);

      model.trigger('sync:start');
    }
  }, options);

  if (!params.url) {
    params.url = getUrl(model) || urlError();
  }

  // Ensure that we have the appropriate request data.
  if (!params.data && model && (method == 'create' || method == 'update')) {
    params.contentType = 'application/json';
    var data = {}

    if(model.paramRoot) {
      data[model.paramRoot] = model.toJSON();
    } else {
      data = model.toJSON();
    }

    params.data = JSON.stringify(data)
  }

  // Don't process data on a non-GET request.
  if (params.type !== 'GET') {
    params.processData = false;
```

```
    }

    // Trigger the sync end event
    var complete = options.complete;
    options.complete = function(jqXHR, textStatus) {
      model.trigger('sync:end');
      if (complete) complete(jqXHR, textStatus);
    };

    // Make the request.
    return $.ajax(params);
  }

}).call(this);
```

说实在的，我不指望你能搞懂上述代码，特别是在我们还没有开始讲模型之前，不过相信我，它对我们后续的代码会很有帮助。因此，只要知道它能帮助我们并感激它就行了。要使用上述代码，首先需要将上述代码放到 assets 目录下的 backbone_sync.js 文件中，然后在 assets/application.coffee 文件中引入 backbone_sync.js 文件。

例（源代码：app.2/assets/application.coffee）

```
#= require "backbone_sync"
#= require "templates"
#= require_tree "models"
```

好了，所有的准备工作都做好了，让我们开始用 Backbone 来构建我们的客户端吧！

12.3 编写 Todo 模型与集合

首先来看的是应用的第一部分——Todo 模型，该模型代表了通过 API 获取到的单个待办事宜。

在 assets 目录中，我们来创建一个叫 models 的新文件夹。该文件夹用来存放 Todo 模型以及 Todos 集合。

例（源代码：app.2/assets/models/todo.coffee）

```
# The Todo model for the Backbone client:
class @Todo extends Backbone.Model
  # namespace JSON under 'todo' see backbone_sync.js
  paramRoot: 'todo'

  # Build the url, appending _id if it exists:
  url: ->
    u = "/api/todos"
    u += "/#{@get("_id")}" unless @isNew()
    return u

  # The default Backbone isNew function looks for 'id',
```

```
  # Mongoose returns "_id", so we need to update it accordingly:
  isNew: ->
    !@get("_id")?
```

写 Backbone 模型时，很重要的一点是模型需要继承自 Backbone.Model 类；否则，无法获得它提供的功能。

因为我们用了 backbone_sync.js，所以需要在数据从服务器端返回时，为其设置一个命名空间的名字。这里，我们通过将 paramRoot 设置为 todo 来实现。

接下来，需要告诉 Backbone，模型要使用哪个 URL 来和 API 通信。为此，我们创建了一个 url 函数。Backbone 会自动查找该函数，并获取 API 所在的位置。当有新的 Todo 对象产生时，该对象是没有 ID 的，所以我们这里只当对象不为新对象时才将 ID 追加到 URL 上。Backbone 内置的 isNew 函数会根据对象是否为“新对象”来对应地返回 true 或者 false。

> **提示**：要想获取 Backbone 对象上的属性，例如：title 或者_id，就必须得使用 get 函数。这是因为 Backbone 模型上的所有属性都存储在一个叫 attributes 的变量中，这样做是为了避免 Backbone 的属性和函数与你设置的属性和函数发生冲突。

Backbone 中的 isNew 函数会查看对象是否有 id 属性，如果存在，则认为该对象不是新对象。遗憾的是，MongoDB[①]返回的不是 id 属性，而是_id 属性。因此，我们需要重写 isNew 函数来正确判断对象是否为新对象。

> **提示**：和 isNew 函数类似，因为 MongoDB 使用的是_id 而不是 id，所以我们不得不重写 url 函数。如果有 id 属性的话，我们可以将 Todo 模型的 url 属性（不是函数）设置为 /api/todos，随后，Backbone 就会自动将 id 属性值追加到 url 属性值之后。不过，因为没有 id 属性而只有_id 属性，所以我们不得不重写该函数。

写好 Todo 模型后，再来写相应的集合 Todos。正如我此前提过的，集合就是列表，和数组相似，存储了很多的 Todo 模型。应用中，我们一开始会使用 Todos 通过我们的 API 来获取所有现存的待办事宜。

> **提示**：我个人觉得，用一个单独的类来管理模型集合多少有点麻烦，毕竟我们需要的这部分功能很少。这仅是学习的一部分。另外补充一下，我通常喜欢将集合类定义与模型类定义放在一个文件中，这样查找方便也利于后期修改。本例中，我将这两个类分开定义是为了更方便地展示代码。

Todos 集合类非常简单。

例（源代码：app.2/assets/models/todos.coffee）

```
# The Todos collection for the Backbone client:
class @Todos extends Backbone.Collection
  model: Todo

  url: "/api/todos"
```

① http://www.mongodb.org/

同样，为了得到 Backbone 为集合提供的功能，Todos 类也需要继承自 Backbone.Collection。然后，我们仅需为集合定义两个属性即可。

第一个属性是 model 属性，我们将该属性设置为 Todo。这就等于告诉集合当从服务器获取到数据，或者接收到数据时，应当将数据转化为 Todo 对象。

第二个属性是 url 属性。因为这是集合，所以无需关心 URL 中的 ID。好了，就这么简单。

至此，我们的 Todos 集合就完成了。

接下来需要更新 assets/application.coffee 文件来引入我们刚刚创建的 models 文件夹。

例（源代码：app.2/assets/application.coffee）

```
#= require "backbone_sync"
#= require "templates"
#= require_tree "models"
```

想必你和我一样，忍不住要看看效果如何。好吧，我们来使用 Todos 集合以及 Todo 模型通过 API 获取现有的待办事宜，并用上一章我们写的模板将它们打印出来。

我们可以简单地通过在 assets/application.coffee 文件中添加几行代码就可以实现，就像下面这样。

例（源代码：app.3/assets/application.coffee）

```
#= require "backbone_sync"
#= require "templates"
#= require_tree "models"

$ ->
  template = _.template(Templates.list_item_template)
  todos = new Todos()
  todos.fetch
    success: ->
      todos.forEach (todo) ->
        $('#todos').append("<li>#{template(todo.toJSON())}</li>")
```

在 DOM 载入完成之后，我们创建了一个新的模板实例，在获取到待办事宜之后，用该模板来将每个待办事宜渲染出来。

随后，我们创建了一个新的 Todos 集合的实例，并将其赋值给了名为 todos 的变量。

有了 Todos 集合实例之后，我们就可以调用 fetch 函数了。fetch 函数会使用我们此前在 Todos 集合中设置的 url 属性与服务器端通信，来获取待办事宜列表。获取成功之后，它会调用我们传递给 fetch 函数的 success 回调函数。

success 回调函数执行时，会调用 todos 对象上的 forEach 函数迭代从服务器端获取的 Todo 模型列表。然后，用模板将每一个待办事宜渲染到屏幕上。

重新载入应用后，就应当能够看到现有的待办事宜都很好地打印在屏幕上了。目前还不能更新或销毁待办事宜。我们会在稍后实现此功能。下一节中，就会开始写第一个 Backbone 视图来替换我们刚刚写的用于展示待办事宜的代码。

12.4 使用视图来罗列待办事宜

虽然我们此前写的代码可以工作，不过它绝对还可以重构得更整洁、更灵活。用 Backbone.View 类就可以实现。那就让我们来用 Backbone.View 类将此前的代码重构得更加整洁吧。

首先，在 assets 文件夹下创建一个 views 文件夹，用它来存放所有的视图文件。我们在该文件夹下创建一个 todo_list_view.coffee 文件，并填入下面的内容。

例（源代码：app.4/assets/views/todo_list_view.coffee）

```
# The 'main' Backbone view of the application
class @TodoListView extends Backbone.View

  el: '#todos'

  initialize: ->
    @template = _.template(Templates.list_item_template)
    @collection.bind("reset", @render)
    @collection.fetch()

  render: =>
    @collection.forEach (todo) =>
      $(@el).append("<li>#{@template(todo.toJSON())}</li>")
```

上述这段代码做了什么呢？问得好。首先，我们创建了一个新类——TodoListView，并让它继承自 Backbone.View。这样就可以使用它提供的一些有用的函数和特性了，这些函数和特性在本章后续部分都会使用到。

> **提示**：注意，和 Todo 类以及 Todos 类一样，我们定义 TodoListView 类时加了@符号前缀。CoffeeScript 为每个.coffee 文件创建了自动包装器函数，有了@就可以在自动包装器函数外部访问到该类。如果没有@的话，就无法在其他.coffee 文件中访问到该类。

接下来，我们告诉视图要将视图绑定在#todos 元素上，可以通过设置 el 属性来实现。如果不这么做的话，Backbone 会为 el 属性创建一个新的 div 对象，你得手动将该元素添加到页面中。后续会看到相关的实践。

继续来看 initialize 函数，它很特别，会在视图对象实例初始化后被 Backbone 调用。千万别在视图类中写 constructor 函数，这样可能会将所有 Backbone 提供的有用的函数和特性都重写掉。如果要在视图初始化时做些事情，那么就用 initialize 函数。

碰巧，我们需要在 TodoListView 类初始化时做些事情。特别是，这些事情和

@collection 对象相关。你可能要问的第一个问题就是：那个变量是哪里来的？Backbone 有些“神奇”的变量和属性，@collection 和@model 就是其中两个。一会儿就能看到，当创建 TodoListView 类的实例时，我们会将一个包含 collection 键、new Todos()值的对象传递给它。该对象会被赋值给@collection 对象，这样在 TodoListView 中，就可以获取 Todos 集合了。等下看到对 application.coffee 文件的必要修改时，应该会更清楚。

我们要用@collection 对象，也就是 Todos 集合做什么呢？首先，调用了 bind 函数，告诉 Backbone，一旦集合触发了 reset 事件就调用 TodoListView 实例上的@render 函数。

集合对象是如何触发 reset 事件的呢？其中一种方式，同时也可能是 Backbone 中最常见的一种方式，就是通过 fetch 函数触发。在调用 fetch 函数时，会通过 API 获取所有的待办事宜列表，正如我们此前看到的一样。因为我们需要这些待办事宜，所以在 initialize 函数的最后一行调用 fetch 函数。这样，调用 fetch 函数继而会触发 reset 事件，然后会调用@render 函数。

@render 函数用来将集合中的待办事宜列表打印到页面上。@render 和我们此前在 application.coffee 中将每个待办事宜渲染到屏幕上的代码没有太大不同。最大的不同就是我们直接使用了@collection 对象上的 forEach 函数，没有调用 success 回调函数。另外一个不同就是我们不再需要直接引用#todos 元素，而是使用@el 即可。使用@el 而不直接使用元素的名字对重构代码很有好处，我们只需修改@el 的值即可，无需对其他代码进行更改。

提示：@render 函数通过使用=>语法而非->来申明函数，是为了当 reset 事件被触发而调用@render 函数时，还能正确获取上下文。如果用->的话，运行这部分代码可能会得到类似于 TypeError: 'undefined' is not an object (evaluating'this collection.forEach')的错误，这是因为@render 已经无法获取@collection 对象了。如果真要使用->语法的话，我们还得在 initialize 函数中，通过 Underscore 库中的 bindAll 函数将@render 函数上下文绑定到视图实例中。_.bindAll(@, "render")。我更喜欢直接使用=>语法。

最后剩下的就是简化 application.coffee 来使用新的 TodoListView 类。

例（源代码：app.4/assets/application.coffee）

```
#= require "backbone_sync"
#= require "templates"
#= require_tree "models"
#= require_tree "views"

$ ->
  # Start Backbone.js App:
  new TodoListView(collection: new Todos())
```

如上述代码所示，我们首先要确保引入了 views 目录，这样才能使用其中的视图。之后，当页面载入完成之后，要做的就是创建一个 TodoListView 的实例，并将该实例传递给 Todos 集合。至此，application.coffee 文件就彻底完成了。

12.5　创建新的待办事宜

可以将现有的待办事宜很好地展示之后，我们继续来实现通过表单创建新待办事宜的功能。要完成这部分工作，需要一个视图来管理该表单，当用户输入待办事宜内容之后按下回车键时处理相关事宜，这样我们就能将新的待办事宜保存到服务器端，然后显示在屏幕上。

我们所需的 NewTodoView 类如下所示。

例（源代码：app.5/assets/views/new_todo_view.coffee）

```
# The view to handle creating new Todos:
class @NewTodoView extends Backbone.View

  el: '#new_todo'

  events:
    'keypress .todo_title': 'handleKeypress'

  initialize: ->
    @collection.bind("add", @resetForm)
    @$('.todo_title').focus()

  handleKeypress: (e) =>
    if e.keyCode is 13
      @saveModel(e)

  resetForm: (todo) =>
    @$('.todo_title').val("")

  saveModel: (e) =>
    e?.preventDefault()
    model = new Todo()
    model.save {title: @$('.todo_title').val()},
      success: =>
        @collection.add(model)
      error: (model, error) =>
        if error.responseText?
          error = JSON.parse(error.responseText)
        alert error.message
```

这比我们刚刚写的 TodoListView 类稍微长了点，不过其中大部分代码都是 saveModel 函数，现在你对这个函数可能还有点陌生。不过，我们很快就会对其做解释。

NewTodoView 需要将自己和页面上的#new_todo 元素绑定起来，所以我们可以通过设置 el 属性来实现。

接下来，我们得告诉 NewTodoView 类去监听特定的事件，并当其事件触发时，做出相应的反应。在 Backbone 中，通过 events 对象属性可以轻松地实现事件绑定。不过，使用 events

属性来进行事件绑定多少有点儿怪。要创建的事件对象的键是一个组合键。其中第一部分是在等待的事件，诸如 `click`、`submit`、`keypress` 等，随后，另外一部分是用来监听事件的 CSS 选择器[①]。创建的事件对象的值是一个函数，当 CSS 选择器匹配到的元素上指定的事件被触发时，就会调用该函数。本例中，我们监听了.todo_title 匹配到的元素上的 `keypress` 事件，当事件触发时，调用 `handleKeypress` 函数。

> **提示：** 在 Backbone 中，关于事件绑定有两件非常重要的事情要注意。第一，CSS 选择器的匹配范围限定在 el 属性指定的元素内。第二，传递的是字符串形式的函数名，并非像此前我们在集合中做绑定时是对函数的引用。我不知道为什么这两种方式不统一，不过，事实就是如此。如果代码出了问题，记得要检查上述两点。

在 `initialize` 函数中，我们将 `resetForm` 函数绑定到`@collection` 对象的 add 事件上，该对象会在创建 `NewTodoView` 类的实例时传递进来。紧接着在 `saveModel` 函数中，当通过 API 成功创建了新的待办事宜后，我们就将其添加到`@collection` 中。这会触发 `add` 事件，然后会调用 `resetForm` 函数。如上述代码所示，`resetForm` 函数会将表单恢复到用户没有输入前的初始状态。

另外，在 `initialize` 函数中，当页面载入完成之后，我们还将焦点聚焦到了表单中的`.todo_title` 元素上。这里我们可以使用 `Backbone.View` 类上的特殊函数——@$函数，它允许直接在绑定到视图的`@el` 元素作用域下写 jQuery CSS 选择器。没有这个特殊函数的话，我们就得写类似`$('#new_todo .todo_title')`这样的代码才能获取到相同的元素。

`handleKeypress` 函数对你来说应当很熟悉了。我们检查用户按下的是否是回车键；如果是，就调用`@saveModel` 函数来处理通过 API 保存数据模型的操作。

> **提示：** 我也可以把保存模型部分的代码直接写在 `handleKeypress` 函数中，之所以不这么做，是因为，当我们还需要添加保存按钮时，就只需要通过 events 属性将按钮和 `saveModel` 函数绑定在一起，不需要重复这部分代码。

`saveModel` 和其他我们写过的处理类似事情的函数很像。首先，我们创建了一个 `Todo` 类的实例。然后，我们收集了一系列的属性，本例中就只是标题，并将其传递给 `save` 函数，同时绑定 `success` 回调函数和 `error` 回调函数。

> **提示：** 为什么要写 `e?.preventDefault()`而不是 `e.preventDefault()`呢？这是为了避免有的时候我们只调用此函数而不传递 `event` 对象而导致的错误。通过使用 CoffeeScript 中有关存在的操作符，我们可以确保 `preventDefault` 函数只在有事件对象时才会调用。这是个很好的习惯。

---

① 该选择器指定要监听此事件的 DOM 元素。——译者注

这里真正神奇的地方发生在，当通过 API 成功将待办事宜保存到数据库中之后执行的 success 回调函数中。该函数中，我们调用了@collection 对象上的 add 函数，并将新创建的待办事宜传递给它。这样一来，resetForm 函数就会被调用，因为我们告诉过@collection 当 add 事件触发时，就调用该函数。

如果现在重启应用看效果的话，什么都不会发生。因为我们还没有创建 NewTodoView 类的实例。要完成此项工作，让我们来对 TodoListView 类做一点小修改，让它不仅能够创建新的 NewTodoView 实例，还能在有新的待办事宜添加到@collection 中时，将它们显示到页面上。

例（源代码：app.5/assets/views/todo_list_view.coffee）

```
# The 'main' Backbone view of the application
class @TodoListView extends Backbone.View

  el: '#todos'

  initialize: ->
    @template = _.template(Templates.list_item_template)
    @collection.bind("reset", @render)
    @collection.fetch()
    @collection.bind("add", @renderAdded)
    new NewTodoView(collection: @collection)

  render: =>
    @collection.forEach (todo) =>
      $(@el).append("<li>#{@template(todo.toJSON())}</li>")

  renderAdded: (todo) =>
    $("#new_todo").after("<li>#{@template(todo.toJSON())}</li>")
```

在 TodoListView 类的 initialize 函数中，我们加了两行代码。第一行代码是给@collection 对象上的 add 事件添加另外一个监听器，这次是触发 TodoListView 类的 renderAdded 函数。

在 TodoListView 类的 initialize 函数中添加的第二行代码是创建一个 NewTodoView 的实例，并将@collection 传递给它。

当新的待办事宜添加到@collection 对象中时，会触发 TodoListView 类中的 renderAdded 函数，同时还会将新创建的待办事宜传递给该函数。有了这个新的待办事宜，我们就可以很容易地将它添加到待办事宜列表中并将其显示在页面上。

> **提示**：我们也可以把这部分代码都写在 render 函数中，那就不需要 renderAdded 函数了。这里之所以没有这么做有几个原因。首先，如果只是添加了新的待办事宜就要渲染整个待办事宜列表，显然更耗时。其次，为了避免重复，我们还得在 render 函数中添加额外的逻辑将页面上现有的待办事宜列表清除。

12.6　每个待办事宜一个视图

从第 11 章中展示了所有我们写的 jQuery 代码开始，就预示着我们的应用已经完成了绝大部分内容了，不过，尚未完全实现。我们可以创建新的待办事宜并与现存的待办事宜一起显示在页面上，但还不能编辑或删除现有的待办事宜。在实现这部分功能之前，需要对已写的代码进行重构，使其更为整洁。我们需要一个视图来管理页面上的单个待办事宜，这样就能监听它们的变化了，比方说，有人编辑了标题、将状态标记为“完成”或者要删除待办事宜。

现在，我们更新代码来为页面上每一个待办事宜使用新的视图——TodoListItemView。好，我们来创建这个类：

例（源代码：app.6/assets/views/todo_list_item_view.coffee）

```
# The view for each todo in the list:
class @TodoListItemView extends Backbone.View

  tagName: 'li'

  initialize: ->
    @template = _.template(Templates.list_item_template)
    @render()

  render: =>
    $(@el).html(@template(@model.toJSON()))
    if @model.get('state') is "completed"
      @$('.todo_state').attr('checked', true)
      @$('label.active').removeClass('active')
      @$('.todo_title').addClass('completed').attr('disabled', true)
    return @
```

TodoListItemView 类不需要映射到页面上现有的元素，所以无须设置 el 属性。Backbone.View 类中的 el 默认是一个 div 标签。不过在本例中，我们想用 li 标签。因此，我们将 tagName 属性设置为 li 标签。

在 TodoListItemView 类的 initialize 函数中，我们定义了@template 变量，该变量存有将待办事宜渲染到屏幕上所需的模板。随后，我们调用了@render 函数。

render 函数会设置@el（也就是本例中的 li 元素）的 HTML 内容为使用了@model 对象数据的模板。那么，@model 是哪里来的呢？@model 对象是在初始化 TodoListItemView 类时传入进来的，稍后就会看到了。

render 函数中，在渲染完模板之后，如果待办事宜的状态为“完成”，就需要更新其 HTML，为其添加对应的样式。

最后，render 函数返回 TodoListItemView 类的实例。并非一定要返回该对象，只是在 Backbone 中这是种约定，因为这样就可以很容易地在一个对象上实现链式调用。

写完 TodoListItemView 之后，我们就可以更新 TodoListView 来使用它了，并且可以删除使用模板渲染的代码了。

例 （源代码：app.6/assets/views/todo_list_view.coffee）

```
# The 'main' Backbone view of the application
class @TodoListView extends Backbone.View

  el: '#todos'

  initialize: ->
    @collection.bind("reset", @render)
    @collection.fetch()
    @collection.bind("add", @renderAdded)
    new NewTodoView(collection: @collection)

  render: =>
    @collection.forEach (todo) =>
      $(@el).append(new TodoListItemView(model: todo).el)

  renderAdded: (todo) =>
    $("#new_todo").after(new TodoListItemView(model: todo).el)
```

在 TodoListView 的 render 和 renderAdded 函数中，我们能通过创建 TodoListItemView 类的实例，获取该类的 el 属性来取代模板的方式。记住，这里 el 属性中的 HTML 片段是在 TodoListItemView 类的 render 函数中定义的。

现在如果重启我们的应用，应该就能看到所有的待办事宜都很好地罗列出来了，并且所有此前被标记为“完成”的待办事宜现在都应该有了对应的 CSS 样式。

12.6.1 从视图层更新和校验模型

现在，我们有了和页面上每个待办事宜相关联的视图——TodoListItemView，有了一个不错的地方来放置监听待办事宜变化并做出相应响应的逻辑。我们先来监听待办事宜的变化。一般待办事宜会发生两种变化。编辑标题和勾选或者取消勾选复选框来改变待办事宜的状态。在下一节会介绍删除待办事宜。

例 （源代码：app.7/assets/views/todo_list_item_view.coffee）

```
# The view for each todo in the list:
class @TodoListItemView extends Backbone.View

  tagName: 'li'

  events:
    'keypress .todo_title': 'handleKeypress'
```

```
    'change .todo_state': 'saveModel'

  initialize: ->
    @template = _.template(Templates.list_item_template)
    @model.bind("change", @render)
    @model.bind("error", @modelSaveFailed)
    @render()

  render: =>
    $(@el).html(@template(@model.toJSON()))
    if @model.get('state') is "completed"
      @$('.todo_state').attr('checked', true)
      @$('label.active').removeClass('active')
      @$('.todo_title').addClass('completed').attr('disabled', true)
    return @

  handleKeypress: (e) =>
    if e.keyCode is 13
      @saveModel(e)

  saveModel: (e) =>
    e?.preventDefault()
    attrs = {title: @$('.todo_title').val()}
    if @$('.todo_state').attr('checked')?
      attrs.state = 'completed'
    else
      attrs.state = 'pending'
    @model.save attrs

  modelSaveFailed: (model, error) =>
    if error.responseText?
      error = JSON.parse(error.responseText)
    alert error.message
    @$('.todo_title').val(@model.get('title'))
```

首先要做的就是添加一些新事件。第一，给.todo_title 匹配的元素添加 keypresss 事件。就和 NewTodoView 类中的一样，事件触发时，会去调用 handleKeypress 函数，检查是否按下了回车键。如果是，就调用 saveModel 函数。除此之外，我们还监听.todo_state 复选框的 change 事件。复选框上发生任何变化都会触发调用 saveModel 函数。

在 initialize 函数中，我们在@model 对象上绑定了两个事件。一个是 change 事件。如果模型发生变化，就会调用 render 函数，确保页面上显示的是最新的待办事宜。同时，当有人勾选了复选框，改变了待办事宜的状态时，也可以很方便地通过添加/移除 CSS 来赋予适当的样式。

我们绑定的另外一个事件是 error 事件。当通过 API 尝试保存待办事宜发生错误时，会触发该事件。随后，会调用 modelSaveFailed 函数，该函数会将错误显示给用户。

最后，我们需要一个 saveModel 函数，因为当用户更新待办事宜时，我们已经告诉了 Backbone 要去调用该函数。saveModel 函数现在不需要多做解释。就是简单地将需要更新的属性传递给 save 函数即可。

> **提示**：在 NewTodoView 类中，我们给 saveModel 函数传递了 success 回调函数和 error 回调函数，但在 TodoListItemView 类中并没有这样做。原因在于我们已经在 @Model 对象上监听了此类事件。而之所以不在 NewTodoView 中使用监听事件的方式，是因为我们会频繁地创建新的 Todo 实例，这样一来我们就得频繁地进行事件绑定。显然，使用回调函数的方式会更加简单。

至此，我们应当可以更新待办事宜的标题和状态，并持久化到服务器端，更新之后也可以正确地渲染出来。

12.6.2 校验

在结束介绍更新待办事宜之前，让我们来为 Todo 类添加一些简单的客户端校验，这样就无须非要等到服务器端再去做校验了。我们尤其关心 title 属性是否为空。

因为我们的 Todo 类是继承自 Backbone.Model 的，这样自然就有了简单的校验系统。该系统的工作方式是这样的：当 Backbone.Model 上的 save 函数被调用时，它就会去检查是否有叫 validate 的函数。如果有，一个包含所有更改的属性的对象就会传递给该函数。如果 validate 函数返回了一个非 null 的值，save 函数就会立刻停止，并将 validate 返回的值返回出去。

我们来给 Todo 模型添加一个 validate 函数。

例 （源代码：app.7/assets/models/todo.coffee）

```
# The Todo model for the Backbone client:
class @Todo extends Backbone.Model
  # namespace JSON under 'todo' see backbone_sync.js
  paramRoot: 'todo'

  # Build the url, appending _id if it exists:
  url: ->
    u = "/api/todos"
    u += "/#{@get("_id")}" unless @isNew()
    return u

  # The default Backbone isNew function looks for 'id',
  # Mongoose returns "_id", so we need to update it accordingly:
  isNew: ->
    !@get("_id")?

  # Validate the model before saving:
  validate: (attrs) ->
    if !attrs.title? or attrs.title.trim() is ""
      return message: "Title can't be blank"
```

如上述代码所示，validate 函数非常简单。用来确保 title 属性存在并且其值不为空字符串。如果该属性不存在，抑或其值为空，我们就返回一个包含 message 键的对象，其中键值为"Title can't be blank"。

好了！重启应用。当给新的或者现有的待办事宜的标题设置为空时，就会得到“Title can't be blank”的提示。至此，上述代码就是校验所需的全部代码，无需再添加任何其他代码了。

12.7　从视图删除模型

最后距离完成我们的应用就只剩下完成删除待办事宜的功能了，这非常容易。我们需要更新 TodoListItemView 并监听删除按钮的 click 事件，当事件触发时，调用相应的函数来删除待办事宜并将其从页面上移除。

例（源代码：final/assets/views/todo_list_item_view.coffee）

```
# The view for each todo in the list:
class @TodoListItemView extends Backbone.View

  tagName: 'li'

  events:
    'keypress .todo_title': 'handleKeypress'
    'change .todo_state': 'saveModel'
    'click .danger': 'destroy'

  initialize: ->
    @template = _.template(Templates.list_item_template)
    @model.bind("change", @render)
    @model.bind("error", @modelSaveFailed)
    @render()

  render: =>
    $(@el).html(@template(@model.toJSON()))
    if @model.get('state') is "completed"
      @$('.todo_state').attr('checked', true)
      @$('label.active').removeClass('active')
      @$('.todo_title').addClass('completed').attr('disabled', true)
      return @

  handleKeypress: (e) =>
    if e.keyCode is 13
      @saveModel(e)

  saveModel: (e) =>
    e?.preventDefault()
    attrs = {title: @$('.todo_title').val()}
    if @$('.todo_state').attr('checked')?
```

```
      attrs.state = 'completed'
    else
      attrs.state = 'pending'
    @model.save attrs

  modelSaveFailed: (model, error) =>
    if error.responseText?
      error = JSON.parse(error.responseText)
    alert error.message
    @$('.todo_title').val(@model.get('title'))

  destroy: (e) =>
    e?.preventDefault()
    if confirm "Are you sure you want to destroy this todo?"
      @model.destroy
        success: =>
          $(@el).remove()
```

另外绑定事件还需要写一个 `destroy` 函数，该函数会通过 API 将待办事宜删除并在删除成功之后，将其从页面上移除。这就是上述代码中 `destroy` 函数的所有功能。出于礼貌，在调用 `Todo` 模型上的 `destroy` 函数之前，我们使用 Backbone 另一个内置的函数，`confirm` 函数，来询问用户是否真的想要删除该待办事宜。

当待办事宜成功地从服务器端删除之后，我们使用 jQuery 及其 `remove` 函数将待办事宜的 `@el` 元素从页面上删除。好了，至此，我们的应用就完成了。

> **提示：** 当模型被删除时，我们还可以监听一些事件。比方说，在待办事宜集合上，我们可以监听 destroy 事件，然后渲染待办事宜列表。我没有这样做的原因依然是不想当添加了新的待办事宜时，去整个重新渲染待办事宜列表。这里要说这些只是告诉你，如果你需要的话，可以来监听这些事件。

12.8 小结

本章，我们废弃了在第 11 章中使用的 jQuery，取而代之的使用了 Backbone.js 框架。这里还要提醒下，如果想将本章和第 11 章的代码都下载下来进行对比，可以从 Github.com[①]上找到本书的所有代码。

本章介绍了 Backbone 的模型和集合。随后，介绍了如何使用视图和事件来管理页面上的元素，以及它们之间是如何交互的。

对于使用 Backbone.js 写高响应、组织良好的应用前端部分，本章只是介绍了它的皮毛而已。建议去找些关于 Backbone 的优秀教程、博文及视频，进行更深入的学习。

① https://github.com/markbates/Programming-In-CoffeeScript